Dwelling Construction

under the 2006
International Building Code®

INTERNATIONAL
CODE COUNCIL®

Dwelling Construction under the 2006 International Building Code

ISBN-978-1-58001-570-7

Cover Design:	Mary Bridges
Cover Art Director:	Dianna Hallmark
Publications Manager:	Mary Lou Luif
Project Editor:	Roger Mensink
Manager of Development:	Hamid Naderi
Typesetting:	Yolanda Nickoley
Illustrator/Interior Design:	Mike Tamai

First Printing: July 2007
Second Printing: November 2007
Third Printing: January 2008

Printed in the United States of America

DWELLING CONSTRUCTION UNDER THE 2006 INTERNATIONAL BUILDING CODE

Preface

This publication is for the convenience of homebuilders, design professionals, building officials and others involved in the design and construction of detached one- and two-family dwellings in conformance with the 2006 edition of the *International Building Code*® (IBC®). The material applies to those communities or states that utilize the IBC rather than the *International Residential Code*® (IRC®) for construction of such residential occupancies. Although many similarities exist between the two codes, and both are dedicated to safeguarding the public health, safety and welfare, there are differences inherent in the development of two separate documents with varying scope and application.

The 2006 *International Building Code* (IBC) meets the need for a modern, comprehensive building code addressing design, construction and installation of building systems through requirements emphasizing performance. It is founded on broad-based principles that make possible the use of new materials and new system designs. The IBC provides that detached one- and two-family dwellings and multiple single-family dwellings (townhouses) not more than three stories above grade plane in height with a separate means of egress and their accessory structures shall comply with the *International Residential Code* (IRC). Because the IBC is not developed with a focus on detached one- and two-family dwellings—whereas the IRC *is* developed with that specific focus—it is recommended that jurisdictions considering the adoption of the IBC for regulation of all occupancies, including one- and two-family dwellings, review the IBC thoroughly for possible major design and construction implications. Local amendments may be necessary through the code adoption process to address certain issues and eliminate the references to the IRC.

The *International Residential Code (*IRC*)* is a comprehensive, stand-alone residential code that establishes minimum regulations for the construction of one- and two-family dwellings and townhouses, including provisions for fire and life safety, structural, energy conservation, mechanical, fuel-gas, plumbing and electrical systems. The IRC incorporates prescriptive provisions for conventional light-frame construction as well as performance criteria that allow the use of new materials and new building designs.

The IBC and IRC are two of the codes in the family of *International Codes*® published by the International Code Council® (ICC®). All are maintained and updated through an open code development process and are available internationally for adoption by the governing authority to provide consistent enforceable regulations for the built environment.

Dwelling Construction under the 2006 International Building Code applies only to the construction of detached one- and two-family dwellings and their accessory buildings utilizing conventional wood frame construction. It intends to serve as a guide, but does not include all applicable requirements of the IBC. The code and the local building department should be consulted for detailed requirements and other methods of construction. Regional design criteria are also outside the limited scope of this publication. Information on requirements related to design for wind, snow, seismic, flood, soil or other conditions are available through the local building department. Energy conservation, decay resistance and termite control requirements also vary by geographic region.

References to section and table numbers, including the parenthetical notes following the various sections of this publication, are to the 2006 edition of the *International Building Code*.

State of California

The 2007 *California Building Code* (CBC), including provisions for one- and two-family dwellings, is based on the 2006 *International Building Code*. The CBC contains numerous amendments to the model code requirements of the IBC. California amendments specific to one- and two-family dwellings are posted on the International Code Council (ICC) website at:

www.iccsafe.org/CADWEL

For further information on California amendments affecting one- and two-family dwellings, contact:

Housing & Community Development
1800 Third Street
Sacramento, CA 95814
(916) 445-9471
www.hcd.ca.gov/codes/shl

Mailing Address:
P.O. Box 1407
Sacramento, CA 95812-1407

TABLE OF CONTENTS

LEGAL REQUIREMENTS

(Note: All tables and figures appear at the end of the text.)

Building Permit

A building permit must be obtained from the building official before beginning construction. Permit application forms are furnished by the building department. (Sec. 105)

Work started before obtaining a permit is subject to an additional fee. (Sec. 108.4)

Information on Plans and Specifications

When applying for a permit, plans and specifications may be required. These plans and specifications must be drawn to scale upon suitable mate

rial and be of sufficient clarity to indicate the location, nature and extent of the work proposed and show in detail that it will be in accordance with the code and all relevant laws, ordinances, rules and regulations. Plans must include a site plan showing to scale the location of the proposed building and of every existing building on the property. A typical site plan is shown in Figure 1. (Secs. 106.1 and 106.2)

When required by statutes or by the building official, construction documents shall be prepared by a registered design professional. (Sec. 106.1)

Permit on Site

A copy of the building permit must be kept on site until the work is completed and a final inspection has been made. (Sec. 105.7)

Inspections

The permit holder or an agent of the permit holder must notify the building official when the work is ready for inspection. In addition to other inspections that may be necessary during the course of the work, the following inspections are required: (Sec 109)

1. **Footing and Foundation Inspection.** To be made after excavations for footings are complete and any required reinforcing steel is in place. For concrete foundations, forms shall be in place prior to inspection. All materials for the foundation shall be on the job, except ready mixed concrete. (Sec. 109.3.1)

2. **Concrete Slab and Under-floor Inspection.** To be made after all in-slab or under-floor reinforcing steel, building service equipment, conduit, piping accessories and other ancillary equipment items are in place but before any concrete is placed or floor sheathing installed, including the subfloor. (Sec. 109.3.2).

3. **Lowest Floor Elevation**. In flood hazard areas, an elevation certificate is required after placement of the lowest floor before proceeding with construction above. (Sec. 109.3.3)

4. **Frame Inspection.** To be made after the roof deck or sheathing, all framing, fire-blocking and bracing are in place; all pipes, chimneys and vents are complete; and the rough electrical, plumbing and heating wires, pipes and ducts are approved. (Sec. 109.3.4)

5. **Lath and Gypsum Board Inspection.** To be made after all lathing and/or gypsum board, interior and exterior, is in place but before any plastering is applied or before gypsum board joints and fasteners are taped and finished. Inspection is not required for gypsum board that is not part of a fire-resistant assembly. (Sec. 109.3.5)

6. **Fire-resistant Penetrations.** To be made to verify protection of joints and penetrations of fire-resistant assemblies before they are concealed. (Sec. 109.3.6)

7. **Energy Efficiency Inspections.** To be made to determine compliance with energy efficiency requirements of Chapter 13 of the IBC. (Sec. 109.3.7)

8. **Final Inspection.** To be made after all work required by the building permit is completed. (Sec. 109.3.10)

Other inspections may also be required by the building official to verify compliance with the provisions of the code and other laws enforced by the code enforcement agency. (Sec. 109.3.8)

The building official will either approve that portion of the construction as completed or will notify the permit holder or an agent of the permit holder wherein the same fails to comply with the code. Portions that do not comply must be corrected and remain uncovered until approved by the building official. (Sec. 109.6)

LOCATION ON THE LOT

Except where zoning regulations are more restrictive, any exterior wall of a dwelling must be at least 10 feet from the property line unless it is of at least one-hour fire-resistant construction. The area of unprotected and protected windows and other exterior wall openings is not limited, provided the distance to the property line is greater than 5 feet. Windows and other openings greater than 3 feet but not more than 5 feet from the property line are limited in area to 25 percent of the wall area.

Windows and other openings are not permitted within 3 feet or less of the property line. (Secs. 602 and 704.8, and Tables 602 and 704.8)

Any exterior wall of a detached garage or other accessory building containing only Group U occupancies must be at least 5 feet from the property line unless it is of at least one-hour fire-resistant construction. Windows and other openings are not permitted within 3 feet or less of the property line. (Secs. 406.1.2 and 704.8, Table 704.8)

One-hour fire-resistant construction includes the following:

1. Exterior plaster over metal or wire lath on the outside and $^5/_8$-inch (15.9 mm) Type X gypsum wallboard on the inside.
2. Three-fourths-inch (19 mm) drop siding over $^1/_2$-inch (12.7 mm) gypsum sheathing on 2-inch by 4-inch (51 mm by 102 mm) wood studs at 16 inches (406 mm) on center on exterior surface with interior surface treatment as specified in Item 1. Gypsum sheathing nailed with $1^3/_4$-inch (44.5 mm) by No. 11 gage by $^7/_{16}$-inch (11.1 mm) head galvanized nails at 8 inches (203 mm) on center. Siding nailed with 7-penny galvanized smooth box nails.

Additional assemblies for fire-resistance-rated wall assemblies are located in Table 720.1(2). (Sec. 720.1)

SEPARATION BETWEEN DWELLING UNITS

Walls and floors separating dwelling units of a two-family dwelling shall not be less than one-hour fire-resistance rated construction. When an NFPA 13 automatic sprinkler system is installed throughout the building, the separation may be reduced to a $^1/_2$-hour fire-resistance rating. (Secs. 419, 708 and 711)

LIGHT, VENTILATION AND SANITATION

Light and Ventilation

Every space intended for human occupancy shall be provided with natural light by means of exterior glazed openings with an area not less than 8 percent of the floor area or shall be provided with artificial light to provide an average illumination of 10 foot-candles (107 lux) over the area of the room at a height of 30 inches (762 mm) above the floor level. (Sec. 1205)

For the purpose of determining natural light requirements, any room may be considered a portion of an adjoining room when one half of the area of the common wall is open and unobstructed and provides an opening of not less than one tenth of the floor area of the interior room or 25 square feet (2.3 m^2), whichever is greater. (Sec. 1205.2.1)

Exterior openings for natural light shall open directly onto a public way, or a yard or court located on the same lot as the building. Required exterior openings are permitted to open into a roofed porch where the porch abuts a public way, yard or court, has a ceiling height of not less than 7 feet (2134 mm), and has a longer side at least 65 percent open and unobstructed. Skylights are not required to open directly onto a public way, yard or court. (Secs. 1205.2.2 and 1206)

Occupied space within a dwelling unit shall be provided with natural ventilation by means of openable exterior openings with an area of not less than 4 percent of the floor area. Rooms containing bathtubs, showers, spas or similar fixtures require mechanical ventilation with an exhaust capacity of 50 cfm intermittent or 20 cfm continuous operation. (Secs. 1203.1 and 1203.4)

In lieu of required exterior openings for natural ventilation, a mechanical ventilation system may be provided. Such a system for living areas shall be capable of providing outdoor air at a rate of 0.35 air changes per hour or 15 cfm per person, whichever is greater. (Sec. 1203.1)

Rooms without openings to the outdoors may obtain natural ventilation from an adjoining room when the opening between spaces has an area of not less than 8 percent of the floor area of the interior room and is not less than 25 square feet (2.3 m^2). The minimum openable area to the outdoors shall be based on the total floor area being ventilated. (Sec. 1203.4.1.1)

Yards or courts adjacent to required openings for natural ventilation must be on the same property as the building. (Sec. 1203.4.3)

Sanitation

Every dwelling unit shall be provided with a kitchen sink and with a bathroom equipped with facilities consisting of a water closet, lavatory and either a bathtub or shower. Each sink, lavatory and bathtub or shower shall be equipped with hot and cold running water necessary for its normal operation. (Secs. 2901 and 2902)

Shower compartments and walls above bathtubs with installed shower heads shall be finished with a smooth, nonabsorbent surface to a height not less than 70 inches (1778 mm) above the drain inlet. (Sec. 1210.3)

INTERIOR SPACE DIMENSIONS

Room Dimensions

Dwelling units shall have at least one room which shall not have less than 120 square feet (13.9 m^2) of floor area. Other habitable rooms except kitchens shall have an area of not less than 70 square feet (6.5 m^2). Kitchens shall have not less than 50 square feet (4.64 m^2) of gross floor area. (Sec. 1208.3)

Habitable rooms other than a kitchen shall not be less than 7 feet (2134 mm) in any dimension. Kitchens shall have a clear passageway of 3 feet (914 mm) between counter fronts and appliances or counter fronts and walls. (Sec. 1208.1)

Ceiling Heights

Habitable spaces and corridors shall have a ceiling height of not less than 7 feet 6 inches (2286 mm). Kitchens, bathrooms, toilet rooms, storage rooms and laundry rooms may have a ceiling height of not less than 7 feet (2134 mm). Exposed beams or girders spaced at 48 inches (1219 mm) or more on center may project not more than 6 inches below the required ceiling height.

If any room in a building has a sloping ceiling, the prescribed ceiling height for the room is required in only one-half the area thereof. No portion of the room measuring less than 5 feet (1524 mm) from the finished floor to the finished ceiling shall be included in any computation of the minimum area thereof.

If any room has a furred ceiling, the prescribed ceiling height is required in two-thirds the area thereof, but in no case shall the height of the furred ceiling be less than 7 feet (2134 mm). (Sec. 1208.2)

FIRE PROTECTION

Smoke Alarms

Dwelling units shall be provided with approved smoke alarms. A smoke alarm shall be installed in each sleeping room and outside of each separate sleeping area in the immediate vicinity of bedrooms. Each story within a dwelling unit, including basements but not including crawl spaces and uninhabitable attics, shall have at least one smoke alarm. In dwelling units with split levels and without an intervening door between the adjacent levels, a smoke alarm installed on the upper level satisfies the requirement for the adjacent lower level, provided that the lower level is less than one full story below the upper level.

Listed smoke alarms complying with UL 217 shall be installed in accordance with the provisions of NFPA 72 and the manufacturer's instructions.

In new construction, required smoke alarms shall receive their primary power from the building wiring when such wiring is served from a commercial source and shall be equipped with a battery backup. The alarm shall emit a signal when the batteries are low. Wiring shall be permanent and without a disconnecting switch other than as required for overcurrent protection.

The smoke alarms shall be interconnected in such a manner that the activation of one alarm will activate all of the alarms. The alarm shall be clearly audible in all bedrooms over background noise levels with all intervening doors closed. (Sec. 907.2.10)

Automatic Sprinkler System

Installation of an automatic sprinkler system for fire protection purposes is required throughout all buildings containing residential occupancies. (Sec. 903.2.7)

MEANS OF EGRESS

Doors

At least one exterior exit door from the dwelling shall have a clear width opening of not less than 32 inches (813 mm) and be not less than 6 feet 8 inches (2032 mm) in height. Other exterior doors shall be not less than 6 feet 4 inches (1930 mm) in height. Door openings within dwellings shall be not less than 6 feet 6 inches (1981 mm) in height. (Secs. 1008.1.1, 1018.2 and 1019.2)

Door Landings

There shall be a floor or landing on each side of all doors. The floor or landing shall be level, except for exterior landings which may have a slope not to exceed $^1/_4$ unit vertical in 12 units horizontal (2 percent slope). A door is permitted to open at the top step of an interior flight of stairs, provided the door does not swing over the top step. Screen doors and storm doors are permitted to swing over stairs or landings.

The landing shall have a width not less than the width of the door and shall have a length measured in the direction of travel of not less than 36 inches (914 mm). (Secs. 1008.1.4, 1008.1.5 and 1009.4)

Thresholds at doorways shall not exceed 0.75 inch (19.1 mm) in height for sliding doors or 0.5 inch (12.7 mm) for other doors. Raised thresholds and floor level changes greater than 0.25 inch (6.4 mm) at doorways shall be beveled. The landing at an exterior doorway is allowed to be not be more than 7.75 inches (197 mm) below the top of the threshold, provided the door, other than an exterior storm or screen door, does not swing over the landing. (Secs. 1008.1.4 and 1008.1.6)

The required means of egress from a room or space in the dwelling shall not pass through the following adjoining or intervening rooms or areas:

1. garages
2. sleeping areas
3. bathrooms or toilet rooms
4. storage rooms, closets or similar spaces

(Sec. 1014.2)

EMERGENCY ESCAPE AND RESCUE OPENINGS

Emergency Escape and Rescue Opening Required

Basements in dwelling units and every sleeping room shall have at least one operable window, exterior door or similar opening approved for emergency escape and rescue that opens directly into a yard, court or public way. The units shall be operable from the inside to provide a full clear opening without the use of keys or tools. Basements with a ceiling height less than 6 feet 8 inches or a floor area not greater than 200 square feet without habitable space do not require emergency escape and rescue openings.

Approved bars, grilles, grates or similar devices are permitted over the prescribed openings, provided the minimum opening size is maintained and such devices are releasable or removable from the inside without the use of a key, tool or force greater than that which is required for normal operation of the escape and rescue opening. When such devices are installed, the code further requires that smoke alarms be installed throughout the building at the locations required for new construction. (Secs. 1026.1 and 1026.4)

Opening Dimensions

Emergency escape and rescue openings shall have a minimum net clear openable area of 5.7 square feet (0.53 m^2). The minimum net clear opening may be reduced to 5 square feet (0.46 m^2) when the sill height of the opening is not more than 44 inches (1118 mm) above or below the finished ground level adjacent to the opening. The net clear dimensions resulting from normal operation of the window or other opening shall be not less than 24 inches (610 mm) in height and 20 inches (508 mm) in width. The bottom of the net clear opening shall be not more than 44 inches (1118 mm) above the floor. (Secs. 1026.2 and 1026.3)

Window Wells

Emergency escape and rescue windows with a finished sill height below the adjacent ground elevation shall have a window well. The clear horizontal dimensions of the window well shall allow the window to be fully opened and provide a minimum accessible net clear opening of 9 square feet (0.84 m^2), with a minimum dimension of 36 inches (914 mm). Window wells with a vertical depth of more than 44 inches (1118 mm) shall be equipped with an approved, permanently affixed ladder or steps that are accessible with the window in the fully open position. The ladder or steps shall not encroach into the required dimensions of the window well by more than 6 inches (152 mm). Ladders or rungs shall have an inside width of at least 12 inches (305 mm), shall project at least 3 inches (76 mm) from the wall and shall be spaced not more than 18 inches (457 mm) on center vertically for the full height of the window well. (Sec. 1026.5)

STAIRS AND RAILINGS

Stairway Width

Stairways shall have a width of not less than 36 inches (914 mm). (Sec. 1009.1)

Headroom

Every stairway shall have headroom clearance of not less than 6 feet 8 inches (2032 mm) measured vertically from a line connecting the edge of the nosings. The minimum headroom clearance shall be the full width of the stairway and continuous to one tread depth beyond the bottom riser. Spiral stairways are permitted a 6-foot 6-inch (1981 mm) headroom clearance.

Rise and Run

The riser height of steps in a stairway shall not be less than 4 inches (102 mm) or greater than 7.75 inches (197 mm), and the tread depth shall not be less than 10 inches (254 mm). The largest tread depth within any flight of stairs shall not exceed the smallest by more than $^3/_8$ inch (9.5 mm). The greatest riser height within any flight of stairs shall not exceed the smallest by more than $^3/_8$ inch (9.5 mm). Solid risers are required for stairways unless the opening between treads does not permit the passage of a sphere with a diameter of 4 inches (102 mm). A nosing not less than 0.75 inch (19.1 mm) but not more than 1.25 inches (32 mm) shall be provided on stairways with solid risers where the tread depth is less than 11 inches (279 mm). A flight of stairs shall not have a vertical rise greater than 12 feet (3658 mm) between floor levels or landings. (See Figure 2.) (Secs. 1009.3 and 1009.6)

Winders

A winder is a tread with nonparallel edges. In residences, the minimum winder tread depth at the walk line is 10 inches (254 mm), and the minimum winder tread depth is 6 inches (152 mm). (Sec. 1009.3)

Stairway Landings

There shall be a floor or landing at the top and bottom of each stairway. Both the landing width and the depth measured in the direction of travel shall be not less than the width of the stairway. The measurement in the direction of travel need not exceed 48 inches (1219 mm). (Sec. 1009.4)

Enclosed Space

Enclosed usable space under stairways shall be protected on the enclosed side with 0.5-inch (12.7 mm) gypsum board. (Sec. 1009.5.3)

Spiral Stairs

Spiral stairs may be used for means of egress in dwellings. The minimum width of a spiral stairway is 26 inches (660 mm). A 7.5-inch (191 mm) minimum clear tread depth must be provided at a point 12 inches (305 mm) from the narrow edge. The risers shall be sufficient to provide a headroom height of 78 inches (1981 mm) minimum, but riser height shall not be more than 9.5 inches (241 mm). (Sec. 1009.8)

Handrails

Stairways shall have a handrail on at least one side. Handrail height shall be uniform and not less than 34 inches (864 mm) or more than 38 inches (965 mm) above the nosing of the treads. Handrails shall be continuous from the top riser to the bottom riser of the stair. A newel post is allowed to interrupt the continuity of a handrail at a landing. The use of a volute, turnout or starting easing is allowed on the lowest tread. Ends shall be returned or shall terminate in newel posts or safety terminals. A change in elevation consisting of a single riser at an entrance or egress door does not require handrails.

Handrails projecting from a wall shall have a space of not less than $1^1/_2$ inches (38 mm) between the wall and the handrail. Projections into the required width of stairways shall not exceed 4.5 inches (114 mm) at or below the handrail height.

Handrails with a circular cross section shall have an outside diameter not less than $1^1/_4$ inches (32 mm) or more than 2 inches (51 mm). If the handrail is not circular, it shall have a perimeter dimension of at least 4 inches (102 mm) and not greater than 6.25 inches (160 mm) with a maximum cross-section dimension of $2^1/_4$ inches (57 mm). The handgrip portion shall have a smooth surface with no sharp corners. (Secs. 1009.10 and 1012)

Guards

All open-sided walking surfaces, stairways, ramps and landings that are located more than 30 inches (762 mm) above the floor or grade below shall be protected by a guard. Guards are also required at the glazed sides of stairways, ramps and landings that are more than 30 inches (762 mm) above the floor or grade below unless the glazing provided meets the strength and attachment requirements as for guards. Screened porches shall be provided with guards where the walking surface is located more than 30 inches (762 mm) above the floor or grade below.

Guards shall not be less than 42 inches (1067 mm) in height. On the open side of stairs, a guard whose top rail also serves as a handrail is permitted to have a height not less than 34 inches (864 mm) and not more than 38 inches (965 mm).

Guards shall have balusters or an ornamental pattern such that a sphere 4 inches (102 mm) in diameter cannot pass through. Openings for required guards on the sides of stairs shall not allow a sphere of $4^3/_8$ inches (111 mm) to pass through. The portion of guards above 34 inches is allowed to have openings such that an 8-inch (203 mm) sphere cannot pass through. (Sec. 1013)

PRIVATE GARAGES AND CARPORTS

An attached garage shall be separated from the dwelling with a minimum $1/_2$-inch (12.7 mm) gypsum board applied to the garage side. Garages shall be separated from any habitable rooms above by not less than a $5/_8$-inch (15.9 mm) Type X gypsum board or equivalent.

The door between the garage and dwelling shall be solid wood, solid core steel, or honeycomb core steel not less than $1^3/_8$ inches (34.9 mm) thick. A labeled door assembly with a minimum fire protection rating of 20 minutes is also permitted. Garages shall not open into a room used for sleeping purposes. Doors shall be self-closing and self-latching. (See Figure 3.) (Sec. 406.1.4)

A carport having no enclosed areas above and entirely open on two or more sides need not have a fire separation between the carport and the dwelling. (Sec. 406.1.4)

FOUNDATIONS, RETAINING WALLS AND DRAINAGE

Foundations

Footings and foundations of masonry, concrete or treated-wood shall be of sufficient size to safely support the loads imposed as determined from the character of the soil. The top and bottom surfaces of footings must be level and changes of elevation accomplished through steps, except the bottom surface is permitted to slope a maximum of 10 percent. Footings shall extend a minimum of 12 inches below undisturbed ground and below the frost line unless specific alternative designs are utilized. Concrete footings supporting walls of light frame construction are allowed to be sized in accordance with Table 1805.4.2. The minimum compressive strength of concrete is 2,500 psi (17.2 MPa) at 28 days. Additional requirements may apply to footings and foundations in Seismic Design Category C, D, E or F. (Secs. 1805.5.5 and 1805.9)

The top of exterior foundations shall be higher than the street or storm drain inlet a minimum of 12 inches (305 mm) plus 2 percent unless an alternate elevation is approved by the building official. The ground immediately adjacent to the foundation shall be sloped away not less than 1:20 (5-percent slope) for a minimum distance of 10 feet (3048 mm). See Figure 5 for foundation clearances from slopes steeper than one unit vertical in three units horizontal (33.3-percent slope). (Secs. 1803 and 1805)

The building official may require a special investigation to verify soil conditions and foundation design. Submittal of a written report may also be required. Under some conditions it will be necessary for a registered design professional to investigate and classify the soils and to prepare the written report. (Sec. 1802)

Dampproofing Foundation Walls

Foundation walls enclosing interior spaces below finished grade shall be dampproofed on the outside surface by approved methods and materials. Waterproofing may be required where a hydrostatic pressure condition exists due to high ground water. (Sec. 1807)

Retaining Walls

Walls retaining earth are required to be designed to prevent overturning, sliding, excessive foundation pressure and water uplift. (Sec. 1806)

Foundation Drainage

Provisions shall be made for subsoil drainage under the floor and around the foundation in accordance with approved methods. (Sec. 1807.4)

MASONRY FIREPLACES AND CHIMNEYS

General

Plans must be submitted showing the details of construction of masonry fireplaces, including clearances from combustible materials. For illustration of typical construction, see Figure 6. (Sec. 2101.3)

Footings

Footings for masonry fireplaces and chimneys shall be concrete or solid masonry at least 12 inches (305 mm) thick and shall extend at least 6 inches (153 mm) beyond the face of the fireplace or foundation wall on all sides. Footings shall be on natural undisturbed earth or engineered fill at least 12 inches (305 mm) below finished grade and below frost depth. (Secs. 2111.2 and 2113.2)

Seismic Reinforcing and Anchorage

In Seismic Design Category D, masonry fireplaces and chimneys shall be reinforced with four No. 4 continuous vertical bars with minimum $1/4$-inch horizontal ties spaced at not more than 18 inches (457 mm) on center. Two horizontal ties are required at each bend in the vertical bars. Fireplaces greater than 40 inches (1016 mm) wide require additional vertical bars. Masonry chimneys in Seismic Design Category D shall be anchored at each floor, ceiling or roof line more than 6 feet (1829 mm) above grade. Anchors shall consist of two $3/16$-inch by 1-inch (4.8 mm by 25.4 mm) straps embedded a minimum of 12 inches (305 mm) into the chimney, hooked around the outer vertical reinforcing bars and extending 6 inches (152 mm) beyond the bend. Each strap shall be fastened to a minimum of four floor joists with two $1/2$-inch (12.7 mm) bolts.

In Seismic Design Category E or F, masonry chimneys shall be reinforced in accordance with the requirements of Sections 2101 through 2108. (Secs. 2111 and 2113)

Firebox

Masonry fireboxes with a lining of firebrick at least 2 inches (51 mm) in thickness shall have back and sidewalls of solid masonry a minimum of 8 inches thick (203 mm), including the lining. The width of joints between firebricks shall not be greater than $1/4$ inch (6.4 mm). The firebox shall have a minimum depth of 20 inches (508 mm). Masonry over a fireplace opening shall be supported by a lintel of noncombustible material with a minimum required bearing of 4 inches (102 mm) at each end. The fireplace throat or damper shall be located a minimum of 8 inches (203 mm) above the top of the fireplace opening. (Sec. 2111)

Hearth and Hearth Extension

Masonry fireplace hearths and hearth extensions shall be concrete or masonry, supported by noncombustible materials, and reinforced to carry their own weight and all imposed loads. The thickness of fireplace hearths shall be not less than 4 inches (102 mm) and hearth extensions not less than 2 inches (51 mm). Hearth extensions shall extend at least 16 inches (406 mm) from the front of, and at least 8 inches (203 mm) beyond each side of, the fireplace opening. Where the fireplace opening is 6 square feet (0.56 m^2) or larger, the hearth extension shall extend at least 20 inches (508 mm) in front of, and at least 12 inches (305 mm) beyond each side of, the fireplace opening. (Secs. 2111.9 and 2111.10)

Combustible Materials

Masonry fireplaces shall maintain a clearance to combustibles of not less than 2 inches (51 mm) from the front and sides, and not less than 4 inches (102 mm) from the back. Masonry chimneys shall have a minimum airspace clearance to combustibles of 2 inches (51 mm). Chimneys located entirely outside the exterior walls of the building shall have a minimum airspace clearance of 1 inch (25 mm). The airspaces shall not be filled, except to provide fireblocking. Allowances are made for combustible trim, sheathing, siding, flooring and drywall, provided they maintain a minimum 12-inch (305 mm) clearance from the firebox or from the inside surface of the nearest chimney flue lining.

Exposed combustible mantels and trim may be placed directly on the masonry fireplace front, provided they are not within 6 inches (152 mm) of the fireplace opening. Combustible material located above and within 12 inches (305 mm) of the fireplace opening shall not project more than $1/8$ inch (3 mm) for each 1-inch (25 mm) clearance from such opening. Combustible materials located along the sides of the fireplace opening that project more than $1^1/_2$ inches (38 mm) from the face of the fireplace shall have an additional clearance equal to the projection. (Secs. 2111.11 and 2113.19)

Airspaces between masonry fireplaces or chimneys and the structure shall be fireblocked with noncombustible material securely fastened in place at each floor and ceiling. (Secs. 2111.13 and 2113.20)

Exterior Air

Masonry fireplaces require an exterior air supply to ensure proper fuel combustion unless the room is mechanically ventilated and controlled. (Sec. 2111.13)

Flues

The minimum net cross-sectional area of the masonry chimney flue for fireplaces shall be determined in accordance with Section 2113.16. The minimum cross-sectional area of a flue serving appliances shall not be smaller in area than the vent connection on the appliance and in accordance with Section 2113.15. (Secs. 2113.15 and 2113.16)

Masonry chimneys shall be lined. The lining material shall be appropriate for the type of appliance connected. Clay flue lining shall be installed in accordance with Section 2113.12. (Secs. 2113.11 and 2113.12)

Height

Every chimney shall extend at least 3 feet (914 mm) above the highest part of the roof through which it passes and at least 2 feet (610 mm) higher than any portion of the building within 10 feet (3048 mm) of the chimney. (Sec. 2113.9)

Loads on Masonry Chimney

A chimney shall not support any structural load other than its own weight unless it is designed and constructed to support additional loads. (Sec. 3113.8)

FRAMING—GENERAL

Conventional Light-frame Construction

Dwellings are permitted to be constructed in accordance with the provisions of Section 2308 for conventional light-frame wood construction subject to certain limitations. Other methods of construction utilizing a satisfactory design are permitted in accordance with other provisions of the code. (Secs. 2308.1 and 2308.2)

When portions or elements of a building of otherwise conventional construction exceed the limits of Section 2308, those portions and the supporting load path must be designed in accordance with accepted engineering practice and other applicable provisions of the code. (Secs. 2308.1.1 and 2308.4)

The AF&PA *Wood Frame Construction Manual for One- and Two-family Dwellings* (WFCM) is a permissible alternative to the conventional construction provisions of Section 2308 subject to the limitations in the WFCM and the code. (Sec. 2308.1).

When of conventional light-frame construction, dwellings are limited to a maximum of three stories above grade and bearing walls shall not exceed a stud height of 10 feet (3048 mm). Required structural design criteria shall not exceed 40 psf (1916 N/m^2) live load for floors, 50 psf (2395 N/m^2) ground snow load, and 100 miles per hour (mph) (44 m/s) (3-second gust) wind speed. Wind speed of 110 mph (48.4 m/s) is permitted for buildings sited in Exposure B. The height of dwellings is limited to two stories in Seismic Design Category C and one story in Seismic Design Category D or E. Conventional light-frame construction shall not be used in irregular portions of structures in Seismic Design Category D or E. See Figure 7 for examples of irregular construction. (Secs. 2308.2 and 2308.12.6)

Foundation Plates or Sills

Wood plates or sills shall be bolted or anchored to the foundation with minimum $^1/_2$-inch diameter (12.7 mm) steel bolts or other approved anchors. Bolts shall be embedded at least 7 inches (178 mm) into the concrete or masonry and shall be spaced not more than 6 feet (1829 mm) apart. There shall be a minimum of two bolts per piece with one bolt located not more than 12 inches (305 mm) or less than 4 inches (102 mm) from each end of the piece. A properly sized nut and washer shall be tightened on each bolt to the plate. (Sec. 2308.6)

Anchors for the sills of braced wall lines shall be spaced at not more than 4 feet (1219 mm) on center for dwellings over two stories in height. (Sec. 2803.3.3) In Seismic Design Category D or E, steel plate washers a minimum of 0.229 inch by 3 inches by 3 inches (5.82 mm by 76 mm by 76 mm) in size are required. The hole in the plate washer is permitted to be diagonally slotted with a width of up to $^3/_{16}$ inch (4.76 mm) larger than the bolt diameter and a slot length not to exceed $1^3/_4$ inches (44 mm), provided a standard cut washer is placed between the plate washer and the nut.

Steel bolts with a minimum nominal diameter of $^5/_8$ inch (15.9 mm) shall be used in Seismic Design Category E. (Sec. 2308.12)

Foundation plates or sills that rest on exterior foundation walls and are less than 8 inches (203 mm) from exposed earth shall be of naturally durable or preservative-treated wood. (Sec. 2304.11)

Protection Against Decay and Termites

Wood structural members exposed to the weather, in contact with the ground, or embedded in concrete that is in contact with the ground shall be naturally durable or preservative-treated wood approved for the location. Posts and columns embedded in concrete footings or earth and supporting permanent structures shall be preservative-treated wood approved for ground contact. Posts or columns supporting permanent structures and supported by a concrete or masonry slab or footing that is in direct contact with the earth shall be of naturally durable or preservative-treated wood unless prescribed clearances above ground are maintained and an impervious membrane is installed.

Where located on concrete slabs placed on earth or attached to the inside of concrete basement walls below grade, wood sleepers, sills or furring shall be naturally durable or preservative-treated. Posts, girders, joists and subfloor located in a crawl space must be naturally durable or treated wood when the floor joist are less than 18 inches (457 mm) or wood girders are less than 12 inches (305 mm) above the exposed ground. Wood framing and wood sheathing that rest on exterior foundation walls and are less than 8 inches (203 mm) from exposed earth, and wood siding with a clearance to the ground of less than 6 inches (152 mm) shall be of naturally durable or preservative-treated wood.

In exterior masonry or concrete walls, a beam pocket shall provide a $^1/_2$-inch (12.7 mm) air space on top, sides and end of a girder unless naturally durable or preservative-treated wood is used. (Sec. 2304.11)

Wood Trusses

Truss design drawings shall be submitted to and approved by the building official prior to installation, and shall also accompany the trusses delivered to the jobsite. The manufacturer shall provide a truss placement diagram identifying the location of each truss. Permanent truss bracing specifications shall be included with the truss design and placement drawings (Sec. 2303.4).

Truss members and components shall not be cut, notched, drilled, spliced or otherwise altered in any way without written approval of a registered design professional. The truss design shall take into account any additional loads such as HVAC or plumbing equipment. (Sec. 2303.4)

FLOOR FRAMING

The allowable spans for floor joists shall be as set forth in Tables 2308.8(1) or 2308.8(2). For other grades and or species, the code references the AF&PA *Span Tables for Joists and Rafters*. Allowable spans for various grades of wood floor sheathing are shown in Tables 2304.7(1), 2304.7(2), 2304.7(3) and 2304.7(4). (Sec. 2308.8)

The ends of each joist shall not have less than $1^1/_2$ inches (38 mm) of bearing on wood or metal, or less than 3 inches (76 mm) on masonry. Joists shall be supported laterally at the ends and at each support. Notches on the ends of joists shall not exceed one-fourth the joist depth. Holes bored in joists shall not be within 2 inches (51 mm) of the top or bottom of the joist, and the diameter of any such hole shall not exceed one-third the depth of the joist. Notches in the top or bottom of joists shall not exceed one-sixth the depth and shall not be located in the middle third of the span. Joists framing into the side of a wood girder shall be supported by joist hangers or on ledger strips not less than 2 inches by 2 inches (51 mm by 51 mm). Field cuts, notches and holes are not permitted in engineered wood products such as trusses and I-joist except where permitted by the manufacturer or in accordance with a specific design by a registered design professional. (Secs. 2308.8.1 and 2308.2)

Bearing partitions parallel to joists shall be supported on beams, doubled joists or walls. Bearing partitions perpendicular to joists shall not be offset from supporting girders or walls more than the joist depth. (Sec. 2308.8.4)

Draftstopping in Floors

In combustible construction of two-family dwellings, draftstopping is required in the concealed space of a floor-ceiling assembly so that the area of the concealed space does not exceed 1,000 square feet (93 m²). Draftstopping materials shall not be less than ¹/₂-inch (12.7 mm) gypsum board, ³/₈-inch (9.5 mm) wood structural panel, ³/₈-inch (9.5 mm) particleboard, 1-inch (25-mm) nominal lumber, cement fiberboard, batts or blankets of mineral wool or glass fiber or other approved materials adequately supported. (Sec. 717.3)

Under-floor Ventilation

Crawl space areas shall be ventilated by an approved mechanical means or by openings into the underfloor area walls. Such openings shall have a net area of not less than 1 square foot for each 150 square feet (0.067 m² for each 10 m²) of underfloor area. Openings shall be located so as to provide cross ventilation. Such openings shall be covered with corrosion-resistant wire mesh with openings not greater than ¹/₈ inch (3.2 mm) or other approved materials with the least dimension of the covering not exceeding ¹/₄ inch (6.4 mm) in dimension. Where moisture that is due to climate and groundwater conditions is not considered excessive, the building official may allow operable louvers and may allow the required net area of vent openings to be reduced to 10 percent of the above, provided the underfloor ground surface area is covered with an approved vapor retarder. Where warranted by climatic conditions, ventilation openings to the outdoors are not required if ventilation openings to the interior are provided. (Sec. 1203.3)

Crawl Space Access

Crawl spaces shall be provided with a minimum of one access opening not less than 18 inches by 24 inches (457 mm by 610 mm). (Sec. 1209.1)

ROOF AND CEILING FRAMING

Tables 2308.10.3(1) through 2308.10.3(6) give the maximum allowable spans for roof rafters. For ceiling joists, see Tables 2308.10.2(1) and 2308.10.2(2). For other grades and or species, the code references the AF&PA *Span Tables for Joists and Rafters*. In areas subject to snowfall, verification shall be obtained from the building official as to the ground snow load or the appropriate rafter table to use.

The allowable span of roof rafters shall be measured along the horizontal projection between supports. Rafters shall be framed directly opposite each other at a ridge board not less than 1-inch nominal thickness and the full depth of the rafter cut.

Roof rafters and trusses shall be supported laterally to prevent rotation and lateral displacement. Ceiling joists and rafter ties shall be fastened to adjacent rafters in accordance with Tables 2308.10.4.1 and 2304.9.1 to provide a continuous rafter tie across the building and shall have a bearing surface of not less than 1¹/₂ inches (38 mm) on the top plate at each end. Where ceiling joists or rafter ties are not provided at a spacing of not more than 4 feet (1219 mm) on center, a ridge beam is required.

Rafters and trusses must be tied to the wall below to resist wind uplift. Uplift loads must be transferred to the foundation using a continuous load path. Rafter or truss to wall connections shall comply with Tables 2304.9.1 and 2308.10.1.

Notching at the ends of rafters or ceiling joists shall not exceed one-fourth the depth. Notches in the top or bottom of the rafter or ceiling joist shall not exceed one-sixth the depth and shall not be located in the middle one-third of the span. Holes bored in rafters or ceiling joists shall not be within 2 inches (51 mm) of the top and bottom and their diameter shall not exceed one-third the depth of the member.

Field cuts, notches and holes are not permitted in engineered wood products such as trusses, glued-laminated lumber and I-joists except where permitted by the manufacturer or in accordance with a specific design by a registered design professional. (Secs. 2303.4.1.7 and 2308.10.7)

Allowable spans in various grades of wood structural panel and lumber roof sheathing are shown in Tables 2304.7(1), 2304.7(2), 2304.7(3) and 2304.7(5). (Roof framing details are illustrated in Figure 13.) (Sec. 2308.10)

Draftstopping in Attics

In combustible construction, draftstopping is required to subdivide attics and concealed roof spaces, such that any horizontal area does not exceed 3,000 square feet (279 m²). Draftstopping materials shall not be less than ¹/₂-inch (12.7 mm) gypsum board, ³/₈-inch (9.5 mm) wood structural panel, ³/₈-inch (9.5 mm) particleboard, 1-inch (25-mm) nominal lumber, cement fiberboard, batts or blankets of mineral wool or glass fiber, or other approved materials adequately supported. (Sec. 717.4)

Attic Ventilation

Attics and enclosed rafter spaces shall have cross ventilation for each separate space. A minimum of 1 inch (25 mm) of airspace is required between the insulation and the roof sheathing. The net free ventilating area shall not be less than ¹/₁₅₀ of the area of the space ventilated, with half of the venting located in the upper portion of the space and the balance provided by soffit vents.

The minimum net free ventilating area may be reduced to ¹/₃₀₀ of the area when an approved vapor retarder is installed on the warm side of the attic insulation. Exterior openings into the attic space shall be covered with approved material with openings a minimum of ¹/₈ inch (3.2 mm) and not exceeding ¹/₄ inch (6.4 mm). (Sec. 1203.2)

Attic Access

An opening not less than 20 inches by 30 inches (559 mm by 762 mm) shall be provided to any attic area having a clear height of over 30 inches (762 mm). A 30-inch (762 mm) minimum clear headroom in the attic space shall be provided at or above the access opening. (Sec. 1209.2)

WALL FRAMING

Framing Details

The size, height and spacing of studs shall be in accordance with Table 2308.9.1, except that Utility grade studs shall not be spaced more than 16 inches (406 mm) on center, nor support more than a roof and ceiling, nor exceed 8 feet (2438 mm) in height for exterior walls and load-bearing walls or 10 feet (3048 mm) for interior nonload-bearing walls.

Studs shall be placed with their wide dimension perpendicular to the wall. Not less than three studs shall be installed at each corner of an exterior wall. The third stud may be omitted through the use of wood spacers or backup cleats of $3/8$-inch-thick (9.5 mm) wood structural panel, $3/8$-inch (9.5 mm) Type 2-M particleboard, 1-inch-thick (25 mm) lumber or other approved devices that will serve as an adequate backing for the attachment of facing materials. Where fire-resistance ratings or shear values are involved, wood spacers, backup cleats or other devices shall not be used unless specifically approved for such use.

Bearing and exterior wall studs shall be capped with double top plates installed to provide overlapping at corners and at intersections with other partitions. End joints in double top plates shall be offset at least 48 inches (1219 mm) and nailed with not less than eight 16d face nails on each side of the joint. A single top plate may be used, provided the plate is adequately tied at joints, corners and intersecting walls by at least the equivalent of 3-inch by 6-inch by 0.036-inch-thick (76 mm by 152 mm by 0.9 mm) galvanized steel that is nailed to each wall or segment of wall by six 8d nails or equivalent, provided the rafters, joists or trusses are centered over the studs with a tolerance of no more than 1 inch (25 mm).

When bearing studs are spaced at 24-inch (610 mm) intervals and top plates are less than two 2-inch by 6-inch (51 mm by 152 mm) or two 3-inch by 4-inch (76 mm by 102 mm) members and when the floor joists, floor trusses or roof trusses that they support are spaced at more then 16-inch (406 mm) intervals, such joists or trusses shall bear within 5 inches (127 mm) of the studs beneath, or a third plate shall be installed.

Interior nonbearing partitions may have studs set with the long dimension parallel to the wall and may be capped with a single top plate installed to provide overlapping at corners and at intersections with other walls and partitions. The plate shall be continuously tied at joints by solid blocking at least 16 inches (406 mm) in length and equal in size to the plate or by $1/2$-inch-by-$1 1/2$-inch (12.7 mm by 38 mm) metal ties with spliced sections fastened with two 16d nails on each side of the joint.

Studs shall have full bearing on a plate or sill not less than 2 inches (51 mm) in thickness having a width not less than that of the wall studs. Wall framing detail is shown in Figure 9. (Sec. 2308.9)

Bracing

Dwellings shall be provided with exterior and interior braced wall lines. Spacing shall not exceed 35 feet (10 668 mm) on center in both the longitudinal and transverse directions in each story. Interior and exterior braced wall line spacing shall not exceed 25 feet (7620 mm) in Seismic Design Category D or E. (Secs. 2308.3 and 2308.12)

See Sections 2308.11 and 2308.12 for additional special requirements specific to the seismic design category assigned to the building.

Connections shall transfer forces from the roofs and floors to braced wall panels and from the braced wall panels in upper stories to the braced wall panels in the story below. (Sec. 2308.3.2)

Braced wall lines shall consist of braced wall panels that meet the requirements for location, type and amount of bracing as shown in Figure 10, specified in Table 2308.9.3(1), and are in line or offset from each other by not more than 4 feet (1219 mm). Braced wall panels shall start at not more than $12 1/2$-feet (3810 mm) from each end of a braced wall line. All braced wall panels shall be clearly indicated on the plans. Construction of braced wall panels shall be by one of the following methods:

1. Nominal 1-inch by 4-inch (25 mm by 102 mm) continuous diagonal braces let into top and bottom plates and intervening studs, placed at an angle not more than 60 degrees (1.0 rad) or less than 45 degrees (0.79 rad) from the horizontal, and attached to the framing in conformance with Table 2304.9.1.

2. Wood boards of $5/8$-inch (16 mm) net minimum thickness applied diagonally on studs spaced not over 24 inches (610 mm) on center.

3. Wood structural panel sheathing with a thickness not less than $5/16$ inch (7.9 mm) for 16-inch (406 mm) stud spacing and not less than $3/8$ inch (9.5 mm) for 24-inch (610 mm) stud spacing in accordance with Tables 2308.9.3(2) and 2308.9.3(3).

4. Fiberboard sheathing panels not less than $1/2$-inch (13 mm) thick, applied vertically or horizontally on studs spaced not over 16 inches (406 mm) on center when installed in accordance with Section 2306.4.4 and Table 2306.4.4.

5. Gypsum board [sheathing-inch (13 mm) thick by 4 feet (1219 mm) wide, wallboard or veneer base] on studs spaced not over 24 inches (610 mm) on center and nailed at 7 inches (178 mm) on center with nails as required by Table 2306.4.5.

6. Particleboard wall sheathing panels where installed in accordance with Table 2308.9.3(4).

7. Portland cement plaster on studs spaced 16 inches (406 mm) on center installed in accordance with Section 2510.

8. Hardboard panel siding when installed in accordance with Section 2303.1.6 and Table 2308.9.3(5).

For cripple wall bracing, see Section 2308.9.4.1.

For Methods 2, 3, 4, 6, 7 and 8, each braced wall panel must be at least 48 inches (1219 mm) in length, covering three stud spaces where studs are 16 inches (406 mm) apart and covering two stud spaces where studs are spaced 24 inches (610 mm) apart.

For Method 5, each braced wall panel must be at least 96 inches (2438 mm) in length when applied to one face of a braced wall panel and 48 inches (1219 mm) when applied to both faces.

All vertical joints of panel sheathing shall occur over studs, and adjacent panel joints shall be nailed to common framing members. Horizontal joints shall occur over blocking equal in size to the studding except where waived by the installation requirements for the specific sheathing materials.

Braced wall panel sole plates shall be nailed to the floor framing, and top plates shall be connected to the framing above in accordance with Section 2308.3.2. Sills shall be bolted to the foundation in accordance with Section 2308.6 except that such anchors shall be spaced at not more than 4 feet (1219 mm) on center for structures over two stories in height. Where joists are perpendicular to braced wall lines above, blocking shall be provided under and in line with the braced wall panels. (Sec. 2308.9.3)

Alternate Braced Wall Panels

Any braced wall panel required by Section 2308.9.3 may be replaced by an alternate braced wall panel constructed in accordance with the following:

1. In one-story buildings, each panel shall have a length of not less than 2 feet 8 inches (813 mm) and a height of not more than 10 feet (3048 mm). Each panel shall be sheathed on one face with $^3/_8$-inch-minimum-thickness (9.5 mm) wood structural panel sheathing nailed with 8d common or galvanized box nails in accordance with Table 2304.9.1 and blocked at all plywood edges. Two anchor bolts installed in accordance with Section 2308.6, shall be provided in each panel. Anchor bolts shall be placed at each panel outside quarter points. Each panel end stud shall have a tie-down device fastened to the foundation, capable of providing an approved uplift capacity of not less than 1,800 pounds (8006 N). The tie-down device shall be installed in accordance with the manufacturer's recommendations. The panels shall be supported directly on a foundation or on floor framing supported directly on a foundation that is continuous across the entire length of the braced wall line. This foundation shall be reinforced with not less than one No. 4 bar top and bottom.

2. In the first story of two-story buildings, each braced wall panel shall be in accordance with Section 2308.9.3.1, Item 1, except that the wood structural panel sheathing shall be provided on both faces, three anchor bolts shall be placed at one-quarter points, and

tie-down device uplift capacity shall not be less than 3,000 pounds (13 344 N). (Sec. 2308.9.3.1)

Alternate Bracing Wall Panel Adjacent to a Door or Window Opening

When used adjacent to a door or window opening with a full-length header with a panel height not exceeding 10 feet (3048 mm), the bracing panel may be reduced to a length of not less than 16 inches (406 mm) in one-story buildings and a length of not less than 24 inches (610 mm) in the first story of two-story buildings, provided all sheathing, fastening and foundation anchorage details meet the requirements of Section 2308.9.3.2. (See Figure 11) (Sec. 2308.9.3.2)

Foundation Cripple Walls

Foundation cripple studs shall not be less in size than the studding above and, when exceeding 4 feet (1219 mm) in height, shall be the size required for an additional story. Cripple walls having a stud height exceeding 14 inches (356 mm) shall be considered a story and braced in accordance with Table 2308.9.3(1) for Seismic Design Category A, B or C. (Sec. 2308.9.4)

In Seismic Design Category D or E, cripple walls having a stud height exceeding 14 inches (356 mm) shall be considered a story and braced in accordance with Table 2308.12.4.

Solid blocking shall be used for cripple walls having a stud height of less than 14 inches (356 mm). (Sec. 2308.9.4)

Headers

Headers are required over openings in exterior bearing walls and interior bearing partitions. Permitted sizes and spans for headers are given in Tables 2308.9.5 and 2308.9.6. Wall studs shall support the ends of the header in accordance with Table 2308.9.5. Each end of the header shall have a length of bearing of not less than $1^1/_2$ inches (38 mm) for the full width. (Secs. 2308.9.5 and 2308.9.6)

Cutting and Notching

In exterior walls and bearing partitions, any wood stud is permitted to be cut or notched to a depth not exceeding 25 percent of its width. Cutting or notching of studs to a depth not greater than 40 percent of the width of the stud is permitted in nonbearing partitions supporting no loads other than the weight of the partition. (Sec. 2308.9.10)

Bored Holes

A hole not greater in diameter than 40 percent of the stud width is permitted to be bored in any wood stud. Bored holes not greater than 60 percent of the width of the stud are permitted in nonbearing partitions or in any wall where each bored stud is doubled, provided not more than two such successive doubled studs are so bored. In no case shall the edge of the bored hole be nearer than $^5/_8$ inch (15.9 mm) to the edge of the stud. (Sec. 2308.9.11)

Fireblocking

In combustible construction, fireblocking shall be installed to cut off all concealed draft openings (both vertical and horizontal) and shall form an effective barrier between floors, between a top story and a roof or attic space. (Sec. 717.1)

Fireblocking shall be provided in the following locations:

1. In concealed spaces of stud walls and partitions, including furred spaces, at the ceiling and floor levels and at 10-foot (3048 mm) intervals horizontally.
2. At all interconnections between concealed vertical stud wall or partition spaces and concealed horizontal spaces created by an assembly of floor joists or trusses, and between concealed vertical and horizontal spaces such as occur at soffits, drop ceilings and cove ceilings.
3. In concealed spaces between stair stringers at the top and bottom of the run.
4. In openings around vents, pipes, ducts, chimneys, fireplaces and similar openings which afford a passage for fire at ceiling and floor levels, with an approved material to resist the free passage of flame and the products of combustion.

Except as provided in Item 4, fireblocking shall consist of 2 inches (51 mm) nominal lumber or two thicknesses of 1-inch (25 mm) nominal lumber with broken lap joints or one thickness of $^{23}/_{32}$-inch (18.3 mm) wood structural panel with joints backed by $^{23}/_{32}$-inch (18.3 mm) wood structural panel or one thickness of $^{3}/_{4}$-inch (19.1 mm) particleboard with joints backed by $^{3}/_{4}$-inch (19.1 mm) particleboard.

Fire blocks may also be of gypsum board, cement asbestos board, mineral fiber, glass fiber or other approved materials securely fastened in place. Loose-fill insulation material shall not be used as a fire block unless specifically fire tested. (Sec. 717.2)

WEATHER PROTECTION

Exterior Walls

Exterior walls shall provide the building with a weather-resistant envelope and prevent the accumulation of water within the wall assembly. A water-resistive barrier installed behind the exterior veneer and a means for draining water that enters the assembly to the exterior are required.

A weather-resistant exterior wall envelope shall not be required over concrete or masonry walls designed in accordance with Chapters 19 and 21. Compliance with the requirements for a means of drainage, water-resistive barrier and flashing shall not be required for an exterior wall envelope that has been demonstrated through testing to resist wind-driven rain. (Sec. 1403.2)

Water-resistive Barrier

A minimum of one layer of No. 15 asphalt felt complying with ASTM D 226 for Type 1 felt or other approved materials shall be applied over studs or sheathing with flashing in such a manner as to provide a continuous water-resistive barrier behind the exterior wall veneer. (Sec. 1404.2)

Flashing

Exterior openings exposed to the weather shall be flashed in such a manner as to make them weatherproof. In addition to exterior door and window assemblies, flashing shall be installed at penetrations and terminations of exterior wall assemblies, exterior wall intersections with roofs, chimneys, porches, decks, balconies and similar projections and at built-in gutters and similar locations where moisture could enter the wall. (Sec. 1405.3)

Exterior Wall Coverings

Exterior walls shall provide weather protection for the building. The materials of the minimum nominal thickness specified in Table 1405.2 shall be acceptable as approved weather coverings. Except as otherwise noted, the minimum thicknesses of the following exterior wall coverings are based on a maximum stud spacing of 16 inches (406 mm) on center. (Sec. 1405.2)

Wood Siding. Solid wood siding shall have a minimum nominal thickness of $^{1}/_{2}$ inch unless placed over sheathing.

Plywood. Where plywood siding is used for covering the exterior of outside walls, it shall be not less than $^{3}/_{8}$ inch (9.5 mm) thick. Plywood panel siding shall be installed in accordance with Table 2308.9.3(2). (Sec. 1405.2)

Hardboard. When hardboard siding is used for covering the outside of exterior walls, it shall be a minimum nominal thickness of $^{3}/_{8}$ inch and installed in accordance with Table 2308.9.3(5). Lap siding shall be installed horizontally and applied to sheathed or unsheathed walls.

Square-edged nongrooved panels and shiplap grooved or nongrooved siding shall be applied vertically to sheathed or unsheathed walls. Siding that is grooved shall not be less than $^{1}/_{4}$ inch (6.4 mm) thick in the groove. (Sec. 1405.2)

Vinyl siding. Vinyl siding complying with ASTM D 3679 shall be permitted on exterior walls of dwellings located in areas where the basic wind speed does not exceed 100 miles per hour (45 m/s) and the building height is less than or equal to 40 feet (12 192 mm) in Exposure C. The siding shall be applied over sheathing or materials listed in Section 2304.6, shall conform to the water-resistive barrier requirements in Section 1403 and shall be installed in accordance with approved manufacturer's instructions. Unless otherwise specified in the approved manufacturer's instructions, nails used to fasten the siding and accessories shall have a minimum 0.313-inch (7.9 mm) head diameter and 0.125-inch (3.18 mm) shank diameter. The nails shall be corrosion resistant and shall be long enough to penetrate the studs or nailing strip at least 0.75 inch (19 mm). Where the siding is installed horizontally, the fastener spacing shall not exceed 16 inches (406 mm) horizontally and 12 inches (305 mm) vertically. Where the siding is installed vertically, the fastener spacing shall not exceed 12 inches (305 mm) horizontally and 12 inches (305 mm) vertically. (Sec. 1405.13)

Fiber cement siding. Fiber cement siding panels shall be installed with the long dimension parallel to framing. Vertical joints shall occur over framing members and shall be sealed with caulking or covered with battens. Horizontal joints shall be flashed with Z-flashing and blocked with solid wood framing. Horizontal lap siding shall be lapped a minimum of $1^1/_4$ inches (32 mm) and shall have the ends sealed with caulking, covered with an H-section joint cover or located over a strip of flashing. (Sec. 1405.17)

Nailing. Siding, weather boarding and wall coverings shall be securely fastened with approved corrosion-resistant fasteners in accordance with the nailing schedule in Table 2304.9.1 or the approved manufacturer's installation instructions. Shingles shall be attached with appropriate standard-shingle nails to furring strips securely nailed to studs, or with approved mechanically bonding nails, except where sheathing is of wood not less than 1-inch (25 mm) nominal thickness or of wood structural panels as specified in Table 2308.9.3(3). (Sec. 1405.16)

Masonry veneer. Anchored and adhered masonry veneer shall comply with the provisions of Sections 1405.5 through 1405.9 and the referenced standards. Masonry materials shall conform to the requirements in Chapter 21. Specific seismic requirements are located in Section 1405.5.2. (Secs. 1405.5 and 1405.9)

Exterior Windows and Doors

Windows and doors shall be installed in accordance with approved manufacturer's instructions. Fastener size and spacing shall be provided in such instructions.

Where the opening of the sill portion of an operable window is located more than 72 inches (1829 mm) above the finished grade or other surface below, the lowest part of the clear opening of the window shall be a minimum of 24 inches (610 mm) above the finished floor surface of the room in which the window is located. Glazing between the floor and a height of 24 inches (610 mm) shall be fixed or have openings such that a 4-inch (102 mm) diameter sphere cannot pass through. Window guards that comply with ASTM F 2006 or F 2090 may be used to satisfy the requirements. (Sec. 1405.12)

GYPSUM BOARD AND PLASTER

Wood Framing

Wood supports for lath or gypsum board, as well as wood stripping or furring, shall not be less than 2 inches (51 mm) nominal thickness in the least dimension. Wood furring strips installed over solid backing are permitted to be not less than nominal 1 inch by 2 inches (25 mm by 51 mm). (Sec. 2504)

Lathing and Plastering

Lathing and plastering materials and accessories shall be marked to indicate conformance to the standards listed in Table 2507.2 and Chapter 35 and stored in such a manner to protect them from the weather. (Sec. 2507)

Lathing and Furring for Cement Plaster (Stucco)

Exterior and interior cement plaster and lathing shall be done with the appropriate materials listed in Table 2507.2 and Chapter 35. Gypsum lath or gypsum wallboard shall not be used as a backing for cement plaster. Gypsum lath or gypsum wallboard is permitted, with a water-resistive barrier as a backing for self-furred metal lath or self-furred wire fabric lath and cement plaster on horizontal supports of ceilings or roof soffits and on interior walls.

Gypsum sheathing is permitted as a backing for metal or wire fabric lath and cement plaster on walls. A water-resistive barrier shall be provided. Water-resistive barriers consisting of one layer of No. 15 asphalt felt complying with ASTM D 226 for Type 1 felt or other approved materials shall be installed as required in Section 1404.2. (Sec. 2510)

Interior Plaster

Plastering with gypsum plaster or cement plaster shall not be less than three coats when applied over metal lath or wire fabric lath and shall not be less than two coats when applied over other bases. (Sec. 2511.1)

Plaster shall not be applied directly to fiber insulation board. Cement plaster shall not be applied directly to gypsum lath, gypsum masonry or gypsum plaster on walls. (Sec, 2511.2)

When installed, grounds shall assure the minimum thickness of plaster as set forth in ASTM C 842 and ASTM C 926. Plaster thickness shall be measured from the face of lath and other bases. (Sec. 2511.3)

Exterior Plaster

Plastering with cement plaster shall not be less than three coats when applied over metal lath or wire fabric lath and shall not be less than two coats when applied over masonry or concrete. If the plaster surface is completely covered by veneer or other facing material, or is completely concealed by another wall, plaster application need be only two coats, provided the total thickness is as set forth in ASTM C 926. (Sec. 2512.1)

Plaster coats shall be protected from freezing for a period of not less than 24 hours after set has occurred. Plaster shall be applied when the ambient temperature is higher than 40°F

(4°C), unless provisions are made to keep cement plaster work above 40°F (4°C) during application and 48 hours thereafter. (Sec. 2512.4)

The second coat shall be brought out to proper thickness, rodded and floated sufficiently rough to provide adequate bond for the finish coat. The second coat shall have no variation greater than $^1/_4$ inch (6.4 mm) in any direction under a 5-foot (1524 mm) straightedge. (Sec. 2512.5)

First and second coats of plaster shall be applied and moist cured as set forth in ASTM C 926 and Table 2512.6. (Sec. 2512.6)

When applied over gypsum backing or directly to unit masonry surfaces, the second coat may be applied as soon as the first coat has attained sufficient hardness. (Sec. 2512.7)

Cement plaster finish coats shall be applied over base coats that have been in place for the time periods set forth in ASTM C 926. The third or finish coat shall be applied with sufficient material and pressure to bond and to cover the brown coat and shall be of sufficient thickness to conceal the brown coat. (Sec. 2512.9)

Gypsum Board

Gypsum board materials and accessories shall be identified to indicate conformance to the appropriate standards listed in Table 2506.2 and Chapter 35. Gypsum board construction shall be of the materials listed in Tables 2506.2 and 2507.2, and installed in conformance to the appropriate standards listed in Tables 2508.1 and 2511.1, and Chapter 35. (Secs. 2506 and 2508)

Gypsum wallboard shall not be installed where it will be exposed directly to the weather, water or high humidity. Gypsum wallboard shall not be installed until weather protection for the installation is provided. (Sec. 2508.2)

All edges and ends of gypsum board shall occur on the framing members, except those edges and ends that are perpendicular to the framing members. All edges and ends of gypsum wallboard shall be in moderate contact except in concealed spaces where fire-resistance-rated construction, shear resistance or diaphragm action is not required. Floating angles are permitted where walls join ceilings except on shear resisting elements or fire-resistance-rated assemblies. Fasteners shall be applied in such a manner as not to fracture the face paper with the fastener head.

Joint Treatment

Gypsum board fire-resistance-rated assemblies shall have joints and fasteners treated. Joint and fastener treatment need not be provided when any of the following conditions occur:

1. Where the gypsum board is to receive a decorative finish such as wood paneling, battens, acoustical finishes or any similar application which would be equivalent to joint treatment.
2. On single-layer systems where joints occur over wood-framing members.
3. Assemblies tested without joint treatment.

(Sec. 2508.4)

Water-resistant gypsum backing board shall not be used over a vapor retarder in shower or bathtub compartments, or in any location subject to exposure to water or continuous high humidity. Water-resistant gypsum backing board shall not be installed on ceilings where frame spacing exceeds 12 inches (305 mm) on center for $^1/_2$-inch-thick (12.7 mm) board and more than 16 inches (406 mm) on center for $^5/_8$-inch-thick (15.9 mm) board. (Sec. 2509.3)

ROOF COVERINGS

General

Roof coverings shall be designed, installed and maintained in accordance with this code and the approved manufacturer's instructions such that the roof covering shall serve to protect the building or structure. Roof-covering materials shall be delivered in packages bearing the manufacturer's identifying marks and approved testing agency labels. (Secs. 1503.1 and 1506.4) Flashing shall be installed in such a manner so as to prevent moisture entering the wall and roof. Flashing shall be installed at wall and roof intersections, at gutters, wherever there is a change in roof slope or direction and around roof openings. Where flashing is of metal, the metal shall be corrosion resistant with a thickness of not less than 0.019 inch (0.483 mm) (No. 26 galvanized sheet). (Sec. 1503.2)

Asphalt Shingles

Each package of asphalt shingles shall bear labeling indicating compliance with ASTM D 3161 or listing of an approved testing agency. Asphalt shingles shall have self-seal strips or be interlocking and comply with ASTM D 225 or ASTM D 3462. For roofs located where the basic wind speed is 110 mph or greater, asphalt shingles shall be tested in accordance with ASTM D 3161, Class F. (Secs. 1507.2.5 and 1504.1.1)

Fasteners for asphalt shingles shall be galvanized, stainless steel, aluminum or copper roofing nails, minimum 12-gage [0.105-inch (2.67 mm)] shank with a minimum $^3/_8$-inch-diameter (9.5 mm) head, of a length to penetrate through the roofing materials and a minimum of $^3/_4$-inch (19.1 mm) into the roof sheathing. Where the roof sheathing is less than $^3/_4$-inch (19.1 mm) thick, the nails shall penetrate through the sheathing. Asphalt shingles shall be fastened according to manufacturer's instructions to solidly sheathed roofs, but not less than four nails per each strip shingle and two nails per each individual shingle shall be used. (Secs. 1507.2.6 and 1507.2.7)

Asphalt shingles shall only be used on roof slopes of two units vertical in 12 units horizontal (17-percent slope) or greater. (Sec. 1507.2.2)

Required underlayment shall conform to ASTM D 226, Type I, ASTM D 4869, Type I, or ASTM D 6757.

For roof slopes from two units vertical in 12 units horizontal (17-percent slope) and up to four units vertical in 12 units horizontal (33-percent slope), underlayment shall be two layers applied with a 19-inch (483 mm) wide starter strip and 19-inch (483 mm) overlap in successive layers. For roof slopes of four units vertical in 12 units horizontal (33-percent slope) or greater, underlayment shall be one layer applied shingle fashion, parallel to and starting from the eave and lapped 2 inches (51 mm), fastened sufficiently to hold in place. Distortions in the underlayment shall not interfere with the ability of the shingles to seal.

Underlayment applied in areas subject to winds greater than 110 mph shall be applied with corrosion-resistant fasteners in accordance with the manufacturer's instructions.

In areas where there has been a history of ice forming along the eaves causing a backup of water, a membrane consisting of two layers of underlayment cemented together or of a self-adhering polymer modified bitumen sheet shall be installed and shall extend from the eave up the roof to a point 24 inches (610 mm) inside the exterior wall line of the building. The ice dam membrane is not required for detached accessory structures that contain no conditioned floor area. (Secs. 1507.2.3 and 1507.2.8)

Wood Shingles

Wood shingles shall be of naturally durable wood and comply with the material standards of Table 1507.8.4. (Sec. 1507.8.4)

Wood shingles may be applied to roofs with solid or spaced sheathing. Spaced sheathing shall be boards not less than nominal 1-inch by 4-inch (25 mm by 102 mm) spaced on centers equal to the weather exposure to coincide with the placement of fasteners. Solid sheathing is required in areas where the average daily temperature in January is 25°F (–4°C) or less, or where there is a possibility of ice forming along the eaves causing a backup of water. (Sec. 1507.8.1)

All wood shingles shall be laid with a side lap of at least $1^1/_2$ inches (38 mm) in adjacent courses, and not in direct alignment in alternate courses. Side spacing between shingles shall be $^1/_4$ to $^3/_8$ inches (6.4 to 9.5 mm). Weather exposures shall not exceed those set forth in Table 1507.8.6. Fasteners for wood shingles shall be corrosion resistant with a minimum penetration of $^3/_4$ inch (19.1 mm) into the sheathing. For sheathing less than $^1/_2$ inch (12.7 mm) in thickness, the fasteners shall extend through the sheathing. Each shingle shall be attached with a minimum of two fasteners. (Secs. 1507.8.5 and 1507.8.6)

Shingles shall not be installed on a roof having a slope less than 3 units vertical in 12 units horizontal (25-percent slope). (Sec. 1507.8.2)

Underlayment shall comply with ASTM D 226, Type I or ASTM D 4869. In areas where there has been a history of ice forming along the eaves causing a backup of water, a membrane consisting of two layers of underlayment cemented together or of a self-adhering polymer modified bitumen sheet shall be installed and shall extend from the eave up the roof to a point 24 inches (610 mm) inside the exterior wall line of the

building. The ice dam membrane is not required for detached accessory structures that contain no conditioned floor area. (Sec. 1507.8.3)

Flashing and counterflashing shall be installed at the juncture of the roof and vertical surfaces in accordance with the manufacturer's installation instructions. Valley flashing shall extend at least 11 inches (279 mm) from the centerline each way and have a splash diverter rib not less than 1 inch (25 mm) high at the flow line formed as part of the flashing. Sections of flashing shall have an end lap of not less than 4 inches (102 mm). The valley flashing shall have a 36-inch-wide (914 mm) underlayment of one layer of type I underlayment or self-adhering polymer-modified bitumen sheet, in addition to other required underlayment. In areas where the average daily temperature in January is 25°F (–4°C) or less or where there is a possibility of ice forming along the eaves causing a backup of water, the valley underlayment, if not self-adhering polymer-modified bitumen sheet, shall be solidly cemented to the roofing underlayment for slopes under 7 units vertical in 12 units horizontal (58-percent slope). (Sec. 1507.8.7)

Wood Shakes

Wood shakes shall comply with the material standards of Table 1507.9.5. (Sec. 1507.9.5) Shakes may be applied to roofs with solid or spaced sheathing. Spaced sheathing shall be boards not less than nominal 1-inch by 4-inch (25 mm by 102 mm) spaced on centers equal to the weather exposure to coincide with the placement of fasteners. Where spaced sheathing is installed at 10 inches (254 mm) on center, additional 1-inch by 4-inch (25 mm by 102 mm) boards shall be installed between the sheathing boards. Solid sheathing is required in areas where the average daily temperature in January is 25°F (–4°C) or less or where there is a possibility of ice forming along the eaves causing a backup of water. (Sec. 1507.9.1)

Shakes shall be laid with a side lap of not less than $1^1/_2$ inches (38 mm) between joints in adjacent courses. Spacing between shakes shall not be less than $^3/_8$ inch (9.5 mm) nor more than $^5/_8$ inch (16 mm), except for preservative-treated wood shakes, which shall have a spacing of not less than $^1/_4$ inch (6.4 mm) nor more than $^3/_8$ inch (9.5 mm). Weather exposures shall not exceed those set forth in Table 1509.7. (Sec. 1507.9.7)

Fasteners for wood shakes shall be corrosion resistant with a minimum penetration of $^3/_4$ inch (19.1 mm) into the sheathing. For sheathing less than $^1/_2$ inch (12.7 mm) in thickness, the fasteners shall extend through the sheathing. Each shake shall be attached with a minimum of two fasteners. (Sec. 1507.9.6)

Shakes shall not be installed on a roof having a slope less than 4 units vertical in 12 units horizontal (33.3 percent slope). (Sec. 1507.9.2)

Interlayment shall comply with ASTM D 226, Type I. Underlayment shall comply with ASTM D 226, Type I or ASTM D 4869. In areas where there has been a history of ice forming along the eaves causing a backup of water, a membrane consisting of two layers of underlayment cemented together or of a self-adhering polymer modified bitumen sheet shall be installed and shall extend from the eave up the roof to a point 24 inches (610 mm) inside the exterior wall line of the building. The ice dam membrane is not required for detached accessory structures that contain no conditioned floor area. (Secs. 1507.9.3 and 1507.9.4)

Flashing and counterflashing shall be installed at the juncture of the roof and vertical surfaces in accordance with the manufacturer's installation instructions. Valley flashing shall extend at least 11 inches (279 mm) from the centerline each way and have a splash diverter rib not less than 1 inch (25 mm) high at the flow line formed as part of the flashing. Sections of flashing shall have an end lap of not less than 4 inches (102 mm). The valley flashing shall have a 36-inch-wide (914 mm) underlayment of one layer of Type I underlayment or self-adhering polymer-modified bitumen sheet, in addition to other required underlayment. In areas where the average daily temperature in January is 25°F (-4°C) or less or where there is a possibility of ice forming along the eaves causing a backup of water, the valley underlayment, if not self-adhering polymer-modified bitumen sheet, shall be solidly cemented to the roofing underlayment for slopes under seven units vertical in 12 units horizontal (58-percent slope). (Sec. 1507.9.8)

SAFETY GLAZING

Glazing, including glass mirrors, in locations subject to human impact, shall be safety glazing meeting the test requirements of CPSC 16 CFR 1201. Plastic glazing, glass block, louvered windows and jalousies shall comply with the other applicable standards. Each pane of safety glazing shall be identified by a manufacturer's designation, including the name of the manufacturer and the safety glazing standard with which it complies. The designation shall be acid etched, sand blasted, ceramic fired, laser etched, embossed or of a type that once applied, cannot be removed without being destroyed. For other than tempered glass, manufacturer's designations are not required, provided the building official approves the use of a certificate, affidavit or other evidence confirming compliance with this code. Tempered spandrel glass is permitted to be identified by the manufacturer with a removable paper designation. (Secs. 2406.1 and 2406.2)

Hazardous locations requiring safety glazing materials include the following:

Glazing in swinging doors except jalousies, in fixed and operable panels of sliding door and bifold door assemblies, and glazing in storm doors.

Glazing in doors and enclosures for hot tubs, whirlpools, saunas, steam rooms, bathtubs and showers, and glazing in any portion of a building wall enclosing these compartments where the bottom exposed edge of the glazing is less than 60 inches (1524 mm) above a standing surface.

Glazing in an individual fixed or operable panel adjacent to a door where the nearest exposed edge of the glazing is within a 24-inch (610 mm) arc of either vertical edge of the door in a closed position and where the bottom exposed edge of the glazing is less than 60 inches (1524 mm) above the walking surface. This provision does not apply when there is an intervening wall between the door and glazing, where access through the door is to a closet 3 feet (914 mm) or less in depth, or to glazing in walls perpendicular to the plane of the door in a closed position, other than the wall towards which the door swings.

Glazing where the individual pane is greater than 9 square feet (0.84 m²), has a bottom edge less than 18 inches (457 mm) above the floor and a top edge greater than 36 inches (914 mm) above the floor, and is adjacent to a walking surface.

Glazing adjacent to stairways, landings and ramps within 36 inches (914 mm) horizontally of a walking surface, when the exposed surface of the glass is less than 60 inches (1524 mm) above the plane of the adjacent walking surface.

Glazing adjacent to stairways within 60 inches (1524 mm) horizontally of the bottom tread of a stairway in any direction, when the exposed surface of the glass is less than 60 inches (1524 mm) above the nose of the tread.

Alternative methods of protection and exceptions to the hazardous location requirements are located in Section 2406.3. Mirrors and other glass panels mounted or hung on a surface that provides a continuous backing support do not require safety glazing. (Sec. 2406.3)

Tables

TABLE 1507.8
WOOD SHINGLE AND SHAKE INSTALLATION

ROOF ITEM	WOOD SHINGLES	WOOD SHAKES
1. Roof slope	Wood shingles shall be installed on slopes of three units vertical in 12 units horizontal (3:12) or greater.	Wood shakes shall be installed on slopes of four units vertical in 12 units horizontal (4:12) or greater.
2. Deck requirement	—	—
Temperate climate	Shingles shall be applied to roofs with solid or spaced sheathing. Where spaced sheathing is used, sheathing boards shall not be less than 1″ × 4″ nominal dimensions and shall be spaced on center equal to the weather exposure to coincide with the placement of fasteners.	Shakes shall be applied to roofs with solid or spaced sheathing. Where spaced sheathing is used, sheathing boards shall not be less than 1″ × 4″ nominal dimensions and shall be spaced on center equal to the weather exposure to coincide with the placement of fasteners. When 1″ × 4″ spaced sheathing is installed at 10 inches, boards must be installed between the sheathing boards.
In areas where the average daily temperature in January is 25°F or less or where there is a possibility of ice forming along the eaves causing a backup of water.	Solid sheathing required.	Solid sheathing is required.
3. Interlayment	No requirements.	Interlayment shall comply with ASTM D 226, Type 1.
4. Underlayment	—	—
Temperate climate	Underlayment shall comply with ASTM D 226, Type 1.	Underlayment shall comply with ASTM D 226, Type 1.
In areas where there is a possibility of ice forming along the eaves causing a backup of water.	An ice shield that consists of at least two layers of underlayment cemented together or of a self-adhering polymer-modified bitumen sheet shall extend from the eave's edge to a point at least 24 inches inside the exterior wall line of the building.	An ice shield that consists of at least two layers of underlayment cemented together or of a self-adhering polymer-modified bitumen sheet shall extend from the eave's edge to a point at least 24 inches inside the exterior wall line of the building.
5. Application	—	—
Attachment	Fasteners for wood shingles shall be corrosion resistant with a minimum penetration of 0.75 inch into the sheathing. For sheathing less than 0.5 inch thick, the fasteners shall extend through the sheathing.	Fasteners for wood shakes shall be corrosion resistant with a minimum penetration of 0.75 inch into the sheathing. For sheathing less than 0.5 inch thick, the fasteners shall extend through the sheathing.
No. of fasteners	Two per shingle.	Two per shake.
Exposure	Weather exposures shall not exceed those set forth in Table 1507.8.6	Weather exposures shall not exceed those set forth in Table 1507.9.7
Method	Shingles shall be laid with a side lap of not less than 1.5 inches between joints in courses, and no two joints in any three adjacent courses shall be in direct alignment. Spacing between shingles shall be 0.25 to 0.375 inch.	Shakes shall be laid with a side lap of not less than 1.5 inches between joints in adjacent courses. Spacing between shakes shall not be less than 0.375 inch or more than 0.625 inch for shakes and tapersawn shakes of naturally durable wood and shall be 0.25 to 0.375 inch for preservative taper sawn shakes.
Flashing	In accordance with Section 1507.8.7.	In accordance with Section 1507.9.8.

For SI: 1 inch = 25.4 mm, °C = [(°F) - 32]/1.8.

TABLE 1507.8.6
WOOD SHINGLE WEATHER EXPOSURE AND ROOF SLOPE

ROOFING MATERIAL	LENGTH (inches)	GRADE	EXPOSURE (inches) 3:12 pitch to < 4:12	EXPOSURE (inches) 4:12 pitch or steeper
Shingles of naturally durable wood	16	No. 1	3.75	5
		No. 2	3.5	4
		No. 3	3	3.5
	18	No. 1	4.25	5.5
		No. 2	4	4.5
		No. 3	3.5	4
	24	No. 1	5.75	7.5
		No. 2	5.5	6.5
		No. 3	5	5.5

For SI: 1 inch = 25.4 mm.

TABLE 1507.9.7
WOOD SHAKE WEATHER EXPOSURE AND ROOF SLOPE

ROOFING MATERIAL	LENGTH (inches)	GRADE	EXPOSURE (inches) 4:12 PITCH OR STEEPER
Shakes of naturally durable wood	18	No. 1	7.5
	24	No. 1	10[a]
Preservative-treated taper sawn shakes of Southern yellow pine	18	No. 1	7.5
	24	No. 1	10
	18	No. 2	5.5
	24	No. 2	7.5
Taper sawn shakes of naturally durable wood	18	No. 1	7.5
	24	No. 1	10
	18	No. 2	5.5
	24	No. 2	7.5

For SI: 1 inch = 25.4 mm.

a. For 24-inch by 0.375-inch handsplit shakes, the maximum exposure is 7.5 inches.

TABLE 1805.4.2
FOOTINGS SUPPORTING WALLS OF LIGHT-FRAME CONSTRUCTION[a, b, c, d, e]

NUMBER OF FLOORS SUPPORTED BY THE FOOTING[f]	WIDTH OF FOOTING (inches)	THICKNESS OF FOOTING (inches)
1	12	6
2	15	6
3	18	8[g]

For SI: 1 inch = 25.4 mm, 1 foot = 304.8 mm.

a. Depth of footings shall be in accordance with Section 1805.2.
b. The ground under the floor is permitted to be excavated to the elevation of the top of the footing.
c. Interior-stud-bearing walls are permitted to be supported by isolated footings. The footing width and length shall be twice the width shown in this table, and footings shall be spaced not more than 6 feet on center.
d. See Section 1908 for additional requirements for footings of structures assigned to Seismic Design Category C, D, E or F.
e. For thickness of foundation walls, see Section 1805.5.
f. Footings are permitted to support a roof in addition to the stipulated number of floors. Footings supporting roof only shall be as required for supporting one floor.
g. Plain concrete footings for Group R-3 occupancies are permitted to be 6 inches thick.

TABLE 2304.9.1
FASTENING SCHEDULE

CONNECTION	FASTENING[a,m]	LOCATION
1. Joist to sill or girder	3 - 8d common ($2^1/_2'' \times 0.131''$) 3 - 3" × 0.131" nails 3 - 3" 14 gage staples	toenail
2. Bridging to joist	2 - 8d common ($2^1/_2''$ ´ 0.131") 2 - 3" × 0.131" nails 2 - 3" 14 gage staples	toenail each end
3. 1" × 6" subfloor or less to each joist	2 - 8d common ($2^1/_2'' \times 0.131''$)	face nail
4. Wider than 1" × 6" subfloor to each joist	3 - 8d common ($2^1/_2'' \times 0.131''$)	face nail
5. 2" subfloor to joist or girder	2 - 16d common ($3^1/_2'' \times 0.162''$)	blind and face nail
6. Sole plate to joist or blocking	16d ($3^1/_2'' \times 0.135''$) at 16" o.c. 3" × 0.131" nails at 8" o.c. 3" 14 gage staples at 12" o.c.	typical face nail
Sole plate to joist or blocking at braced wall panel	3" - 16d ($3^1/_2'' \times 0.135''$) at 16" 4 - 3" × 0.131" nails at 16" 4 - 3" 14 gage staples per 16"	braced wall panels
7. Top plate to stud	2 - 16d common ($3^1/_2'' \times 0.162''$) 3 - 3" × 0.131" nails 3 - 3" 14 gage staples	end nail
8. Stud to sole plate	4 - 8d common ($2^1/_2'' \times 0.131''$) 4 - 3" × 0.131" nails 3 - 3" 14 gage staples	toenail
	2 - 16d common ($3^1/_2'' \times 0.162''$) 3 - 3" × 0.131" nails 3 - 3" 14 gage staples	end nail
9. Double studs	16d ($3^1/_2'' \times 0.135''$) at 24" o.c. 3" × 0.131" nail at 8" o.c. 3" 14 gage staple at 8" o.c.	face nail
10. Double top plates	16d ($3^1/_2'' \times 0.135''$) at 16" o.c. 3" × 0.131" nail at 12" o.c. 3" 14 gage staple at 12" o.c.	typical face nail
Double top plates	8-16d common ($3^1/_2'' \times 0.162''$) 12-3" × 0.131" nails 12-3" 14 gage staples	lap splice
11. Blocking between joists or rafters to top plate	3 - 8d common ($2^1/_2'' \times 0.131''$) 3 - 3" × 0.131" nails 3 - 3" 14 gage staples	toenail
12. Rim joist to top plate	8d ($2^1/_2'' \times 0.131''$) at 6" o.c. 3" × 0.131" nail at 6" o.c. 3" 14 gage staple at 6" o.c.	toenail
13. Top plates, laps and intersections	2 - 16d common ($3^1/_2'' \times 0.162''$) 3 - 3" × 0.131" nails 3 -3" 14 gage staples	face nail
14. Continuous header, two pieces	16d common ($3^1/_2'' \times 0.162''$)	16" o.c. along edge
15. Ceiling joists to plate	3 - 8d common ($2^1/_2'' \times 0.131''$) 5 - 3" × 0.131" nails 5 - 3" 14 gage staples	toenail
16. Continuous header to stud	4 - 8d common ($2^1/_2'' \times 0.131''$)	toenail

(continued)

TABLE 2304.9.1—continued
FASTENING SCHEDULE

CONNECTION	FASTENING[a,m]	LOCATION
17. Ceiling joists, laps over partitions (see Section 2308.10.4.1, Table 2308.10.4.1)	3 - 16d common ($3^1/_2'' \times 0.162''$) minimum, Table 2308.10.4.1 4 - $3'' \times 0.131''$ nails 4 - 3" 14 gage staples	face nail
18. Ceiling joists to parallel rafters (see Section 2308.10.4.1, Table 2308.10.4.1)	3 - 16d common ($3^1/_2'' \times 0.162''$) minimum, Table 2308.10.4.1 4 - $3'' \times 0.131''$ nails 4 - 3" 14 gage staples	face nail
19. Rafter to plate (see Section 2308.10.1, Table 2308.10.1)	3 - 8d common ($2^1/_2'' \times 0.131''$) 3 - $3'' \times 0.131''$ nails 3 - 3" 14 gage staples	toenail
20. 1" diagonal brace to each stud and plate	2 - 8d common ($2^1/_2'' \times 0.131''$) 2 - $3'' \times 0.131''$ nails 3 - 3" 14 gage staples	face nail
21. 1" × 8" sheathing to each bearing	3 - 8d common ($2^1/_2'' \times 0.131''$)	face nail
22. Wider than 1" × 8" sheathing to each bearing	3 - 8d common ($2^1/_2'' \times 0.131''$)	face nail
23. Built-up corner studs	16d common ($3^1/_2'' \times 0.162''$) $3'' \times 0.131''$ nails 3" 14 gage staples	24" o.c. 16" o.c. 16" o.c.
24. Built-up girder and beams	20d common ($4'' \times 0.192''$) 32" o.c. $3'' \times 0.131''$ nail at 24" o.c. 3" 14 gage staple at 24" o.c.	face nail at top and bottom staggered on opposite sides
	2 - 20d common ($4'' \times 0.192''$) 3 - $3'' \times 0.131''$ nails 3 - 3" 14 gage staples	face nail at ends and at each splice
25. 2" planks	16d common ($3^1/_2'' \times 0.162''$)	at each bearing
26. Collar tie to rafter	3 - 10d common ($3'' \times 0.148''$) 4 - $3'' \times 0.131''$ nails 4 - 3" 14 gage staples	face nail
27. Jack rafter to hip	3 - 10d common ($3'' \times 0.148''$) 4 - $3'' \times 0.131''$ nails 4 - 3" 14 gage staples	toenail
	2 - 16d common ($3^1/_2'' \times 0.162''$) 3 - $3'' \times 0.131''$ nails 3 - 3" 14 gage staples	face nail
28. Roof rafter to 2-by ridge beam	2 - 16d common ($3^1/_2'' \times 0.162''$) 3 - $3'' \times 0.131''$ nails 3 - 3" 14 gage staples	toenail
	2 - 16d common ($3^1/_2'' \times 0.162''$) 3 - $3'' \times 0.131''$ nails 3 - 3" 14 gage staples	face nail
29. Joist to band joist	3 - 16d common ($3^1/_2'' \times 0.162''$) 4 - $3'' \times 0.131''$ nails 4 - 3" 14 gage staples	face nail

(continued)

TABLE 2304.9.1—continued
FASTENING SCHEDULE

CONNECTION	FASTENING[a,m]		LOCATION
30. Ledger strip	3 - 16d common ($3^1/_2''$ × 0.162") 4 - 3" x 0.131" nails 4 - 3" 14 gage staples		face nail
31. Wood structural panels and particleboard[b] Subfloor, roof and wall sheathing (to framing)	$^1/_2''$ and less	6d[c,l] $2^3/_8''$ × 0.113" nail[n] $1^3/_4''$ 16 gage[o]	
	$^{19}/_{32}''$ to $^3/_4''$	8d[d] or 6d[e] $2^3/_8''$ × 0.113" nail[p] 2" 16 gage[p]	
	$^7/_8''$ to 1"	8d[c]	
	$1^1/_8''$ to $1^1/_4''$	10d[d] or 8d[d]	
Single Floor (combination subfloor-underlayment to framing)	$^3/_4''$ and less	6d[e]	
	$^7/_8''$ to 1"	8d[e]	
	$1^1/_8''$ to $1^1/_4''$	10d[d] or 8d[e]	
32. Panel siding (to framing)	$^1/_2''$ or less	6d[f]	
	$^5/_8''$	8d[f]	
33. Fiberboard sheathing[g]	$^1/_2''$	No. 11 gage roofing nail[h] 6d common nail (2" × 0.113") No. 16 gage staple[i]	
	$^{25}/_{32}''$	No. 11 gage roofing nail[h] 8d common nail ($2^1/_2''$ × 0.131") No. 16 gage staple[i]	
34. Interior paneling	$^1/_4''$	4d[j]	
	$^3/_8''$	6d[k]	

For SI: 1 inch = 25.4 mm.

a. Common or box nails are permitted to be used except where otherwise stated.

b. Nails spaced at 6 inches on center at edges, 12 inches at intermediate supports except 6 inches at supports where spans are 48 inches or more. For nailing of wood structural panel and particleboard diaphragms and shear walls, refer to Section 2305. Nails for wall sheathing are permitted to be common, box or casing.

c. Common or deformed shank (6d - 2" × 0.113"; 8d - $2^1/_2''$ × 0.131"; 10d - 3" × 0.148").

d. Common (6d - 2" × 0.113"; 8d - $2^1/_2''$ × 0.131"; 10d - 3" × 0.148").

e. Deformed shank (6d - 2" × 0.113"; 8d - $2^1/_2''$ × 0.131"; 10d - 3" × 0.148").

f. Corrosion-resistant siding (6d - $1^7/_8''$ × 0.106"; 8d - $2^3/_8''$ × 0.128") or casing (6d - 2" × 0.099"; 8d - $2^1/_2''$ × 0.113") nail.

g. Fasteners spaced 3 inches on center at exterior edges and 6 inches on center at intermediate supports, when used as structural sheathing. Spacing shall be 6 inches on center on the edges and 12 inches on center at intermediate supports for nonstructural applications.

h. Corrosion-resistant roofing nails with $^7/_{16}$-inch-diameter head and $1^1/_2$-inch length for $^1/_2$-inch sheathing and $1^3/_4$-inch length for $^{25}/_{32}$-inch sheathing.

i. Corrosion-resistant staples with nominal $^7/_{16}$-inch crown and $1^1/_8$-inch length for $^1/_2$-inch sheathing and $1^1/_2$-inch length for $^{25}/_{32}$-inch sheathing. Panel supports at 16 inches (20 inches if strength axis in the long direction of the panel, unless otherwise marked).

j. Casing ($1^1/_2''$ × 0.080") or finish ($1^1/_2''$ × 0.072") nails spaced 6 inches on panel edges, 12 inches at intermediate supports.

k. Panel supports at 24 inches. Casing or finish nails spaced 6 inches on panel edges, 12 inches at intermediate supports.

l. For roof sheathing applications, 8d nails ($2^1/_2''$ × 0.113") are the minimum required for wood structural panels.

m. Staples shall have a minimum crown width of $^7/_{16}$ inch.

n. For roof sheathing applications, fasteners spaced 4 inches on center at edges, 8 inches at intermediate supports.

o. Fasteners spaced 4 inches on center at edges, 8 inches at intermediate supports for subfloor and wall sheathing and 3 inches on center at edges, 6 inches at intermediate supports for roof sheathing.

p. Fasteners spaced 4 inches on center at edges, 8 inches at intermediate supports.

TABLE 2308.8(1)
FLOOR JOIST SPANS FOR COMMON LUMBER SPECIES
(Residential Sleeping Areas, Live Load = 30 psf, L/Δ = 360)

JOIST SPACING (inches)	SPECIES AND GRADE		DEAD LOAD = 10 psf				DEAD LOAD = 20 psf			
			2 x 6	2 x 8	2 x 10	2 x 12	2 x 6	2 x 8	2 x 10	2 x 12
			\(Maximum floor joist spans\)							
			(ft. - in.)	(ft. - in.)	(ft. - in.)	(ft. - in.)	(ft. - in.)	(ft. - in.)	(ft. - in.)	(ft. - in.)
12	Douglas Fir-Larch	SS	12-6	16-6	21-0	25-7	12-6	16-6	21-0	25-7
	Douglas Fir-Larch	#1	12-0	15-10	20-3	24-8	12-0	15-7	19-0	22-0
	Douglas Fir-Larch	#2	11-10	15-7	19-10	23-0	11-6	14-7	17-9	20-7
	Douglas Fir-Larch	#3	9-8	12-4	15-0	17-5	8-8	11-0	13-5	15-7
	Hem-Fir	SS	11-10	15-7	19-10	24-2	11-10	15-7	19-10	24-2
	Hem-Fir	#1	11-7	15-3	19-5	23-7	11-7	15-2	18-6	21-6
	Hem-Fir	#2	11-0	14-6	18-6	22-6	11-0	14-4	17-6	20-4
	Hem-Fir	#3	9-8	12-4	15-0	17-5	8-8	11-0	13-5	15-7
	Southern Pine	SS	12-3	16-2	20-8	25-1	12-3	16-2	20-8	25-1
	Southern Pine	#1	12-0	15-10	20-3	24-8	12-0	15-10	20-3	24-8
	Southern Pine	#2	11-10	15-7	19-10	24-2	11-10	15-7	18-7	21-9
	Southern Pine	#3	10-5	13-3	15-8	18-8	9-4	11-11	14-0	16-8
	Spruce-Pine-Fir	SS	11-7	15-3	19-5	23-7	11-7	15-3	19-5	23-7
	Spruce-Pine-Fir	#1	11-3	14-11	19-0	23-0	11-3	14-7	17-9	20-7
	Spruce-Pine-Fir	#2	11-3	14-11	19-0	23-0	11-3	14-7	17-9	20-7
	Spruce-Pine-Fir	#3	9-8	12-4	15-0	17-5	8-8	11-0	13-5	15-7
16	Douglas Fir-Larch	SS	11-4	15-0	19-1	23-3	11-4	15-0	19-1	23-0
	Douglas Fir-Larch	#1	10-11	14-5	18-5	21-4	10-8	13-6	16-5	19-1
	Douglas Fir-Larch	#2	10-9	14-1	17-2	19-11	9-11	12-7	15-5	17-10
	Douglas Fir-Larch	#3	8-5	10-8	13-0	15-1	7-6	9-6	11-8	13-6
	Hem-Fir	SS	10-9	14-2	18-0	21-11	10-9	14-2	18-0	21-11
	Hem-Fir	#1	10-6	13-10	17-8	20-9	10-4	13-1	16-0	18-7
	Hem-Fir	#2	10-0	13-2	16-10	19-8	9-10	12-5	15-2	17-7
	Hem-Fir	#3	8-5	10-8	13-0	15-1	7-6	9-6	11-8	13-6
	Southern Pine	SS	11-2	14-8	18-9	22-10	11-2	14-8	18-9	22-10
	Southern Pine	#1	10-11	14-5	18-5	22-5	10-11	14-5	17-11	21-4
	Southern Pine	#2	10-9	14-2	18-0	21-1	10-5	13-6	16-1	18-10
	Southern Pine	#3	9-0	11-6	13-7	16-2	8-1	10-3	12-2	14-6
	Spruce-Pine-Fir	SS	10-6	13-10	17-8	21-6	10-6	13-10	17-8	21-4
	Spruce-Pine-Fir	#1	10-3	13-6	17-2	19-11	9-11	12-7	15-5	17-10
	Spruce-Pine-Fir	#2	10-3	13-6	17-2	19-11	9-11	12-7	15-5	17-10
	Spruce-Pine-Fir	#3	8-5	10-8	13-0	15-1	7-6	9-6	11-8	13-6

(continued)

TABLE 2308.8(1)—continued
FLOOR JOIST SPANS FOR COMMON LUMBER SPECIES
(Residential Sleeping Areas, Live Load = 30 psf, L/Δ = 360)

JOIST SPACING (inches)	SPECIES AND GRADE		DEAD LOAD = 10 psf				DEAD LOAD = 20 psf			
			Maximum floor joist spans							
			2 x 6	2 x 8	2 x 10	2 x 12	2 x 6	2 x 8	2 x 10	2 x 12
			(ft. - in.)	(ft. - in.)	(ft. - in.)	(ft. - in.)	(ft. - in.)	(ft. - in.)	(ft. - in.)	(ft. - in.)
19.2	Douglas Fir-Larch	SS	10-8	14-1	18-0	21-10	10-8	14-1	18-0	21-0
	Douglas Fir-Larch	#1	10-4	13-7	16-9	19-6	9-8	12-4	15-0	17-5
	Douglas Fir-Larch	#2	10-1	12-10	15-8	18-3	9-1	11-6	14-1	16-3
	Douglas Fir-Larch	#3	7-8	9-9	11-10	13-9	6-10	8-8	10-7	12-4
	Hem-Fir	SS	10-1	13-4	17-0	20-8	10-1	13-4	17-0	20-7
	Hem-Fir	#1	9-10	13-0	16-4	19-0	9-6	12-0	14-8	17-0
	Hem-Fir	#2	9-5	12-5	15-6	17-1	8-11	11-4	13-10	16-1
	Hem-Fir	#3	7-8	9-9	11-10	13-9	6-10	8-8	10-7	12-4
	Southern Pine	SS	10-6	13-10	17-8	21-6	10-6	13-10	17-8	21-6
	Southern Pine	#1	10-4	13-7	17-4	21-1	10-4	13-7	16-4	19-6
	Southern Pine	#2	10-1	13-4	16-5	19-3	9-6	12-4	14-8	17-2
	Southern Pine	#3	8-3	10-6	12-5	14-9	7-4	9-5	11-1	13-2
	Spruce-Pine-Fir	SS	9-10	13-0	16-7	20-2	9-10	13-0	16-7	19-6
	Spruce-Pine-Fir	#1	9-8	12-9	15-8	18-3	9-1	11-6	14-1	16-3
	Spruce-Pine-Fir	#2	9-8	12-9	15-8	18-3	9-1	11-6	14-1	16-3
	Spruce-Pine-Fir	#3	7-8	9-9	11-10	13-9	6-10	8-8	10-7	12-4
24	Douglas Fir-Larch	SS	9-11	13-1	16-8	20-3	9-11	13-1	16-2	18-9
	Douglas Fir-Larch	#1	9-7	12-4	15-0	17-5	8-8	11-0	13-5	15-7
	Douglas Fir-Larch	#2	9-1	11-6	14-1	16-3	8-1	10-3	12-7	14-7
	Douglas Fir-Larch	#3	6-10	8-8	10-7	12-4	6-2	7-9	9-6	11-0
	Hem-Fir	SS	9-4	12-4	15-9	19-2	9-4	12-4	15-9	18-5
	Hem-Fir	#1	9-2	12-0	14-8	17-0	8-6	10-9	13-1	15-2
	Hem-Fir	#2	8-9	11-4	13-10	16-1	8-0	10-2	12-5	14-4
	Hem-Fir	#3	6-10	8-8	10-7	12-4	6-2	7-9	9-6	11-0
	Southern Pine	SS	9-9	12-10	16-5	19-11	9-9	12-10	16-5	19-11
	Southern Pine	#1	9-7	12-7	16-1	19-6	9-7	12-4	14-7	17-5
	Southern Pine	#2	9-4	12-4	14-8	17-2	8-6	11-0	13-1	15-5
	Southern Pine	#3	7-4	9-5	11-1	13-2	6-7	8-5	9-11	11-10
	Spruce-Pine-Fir	SS	9-2	12-1	15-5	18-9	9-2	12-1	15-0	17-5
	Spruce-Pine-Fir	#1	8-11	11-6	14-1	16-3	8-1	10-3	12-7	14-7
	Spruce-Pine-Fir	#2	8-11	11-6	14-1	16-3	8-1	10-3	12-7	14-7
	Spruce-Pine-Fir	#3	6-10	8-8	10-7	12-4	6-2	7-9	9-6	11-0

Check sources for availability of lumber in lengths greater than 20 feet.
For SI: 1 inch = 25.4 mm, 1 foot = 304.8 mm, 1 pound per square foot = 47.8 N/m².

TABLE 2308.8(2)
FLOOR JOIST SPANS FOR COMMON LUMBER SPECIES
(Residential Living Areas, Live Load = 40 psf, $L/\Delta = 360$)

JOIST SPACING (inches)	SPECIES AND GRADE		DEAD LOAD = 10 psf				DEAD LOAD = 20 psf			
			2 x 6	2 x 8	2 x 10	2 x 12	2 x 6	2 x 8	2 x 10	2 x 12
			(ft. - in.)	(ft. - in.)	(ft. - in.)	(ft. - in.)	(ft. - in.)	(ft. - in.)	(ft. - in.)	(ft. - in.)
						Maximum floor joist spans				
12	Douglas Fir-Larch	SS	11-4	15-0	19-1	23-3	11-4	15-0	19-1	23-3
	Douglas Fir-Larch	#1	10-11	14-5	18-5	22-0	10-11	14-2	17-4	20-1
	Douglas Fir-Larch	#2	10-9	14-2	17-9	20-7	10-6	13-3	16-3	18-10
	Douglas Fir-Larch	#3	8-8	11-0	13-5	15-7	7-11	10-0	12-3	14-3
	Hem-Fir	SS	10-9	14-2	18-0	21-11	10-9	14-2	18-0	21-11
	Hem-Fir	#1	10-6	13-10	17-8	21-6	10-6	13-10	16-11	19-7
	Hem-Fir	#2	10-0	13-2	16-10	20-4	10-0	13-1	16-0	18-6
	Hem-Fir	#3	8-8	11-0	13-5	15-7	7-11	10-0	12-3	14-3
	Southern Pine	SS	11-2	14-8	18-9	22-10	11-2	14-8	18-9	22-10
	Southern Pine	#1	10-11	14-5	18-5	22-5	10-11	14-5	18-5	22-5
	Southern Pine	#2	10-9	14-2	18-0	21-9	10-9	14-2	16-11	19-10
	Southern Pine	#3	9-4	11-11	14-0	16-8	8-6	10-10	12-10	15-3
	Spruce-Pine-Fir	SS	10-6	13-10	17-8	21-6	10-6	13-10	17-8	21-6
	Spruce-Pine-Fir	#1	10-3	13-6	17-3	20-7	10-3	13-3	16-3	18-10
	Spruce-Pine-Fir	#2	10-3	13-6	17-3	20-7	10-3	13-3	16-3	18-10
	Spruce-Pine-Fir	#3	8-8	11-0	13-5	15-7	7-11	10-0	12-3	14-3
16	Douglas Fir-Larch	SS	10-4	13-7	17-4	21-1	10-4	13-7	17-4	21-0
	Douglas Fir-Larch	#1	9-11	13-1	16-5	19-1	9-8	12-4	15-0	17-5
	Douglas Fir-Larch	#2	9-9	12-7	15-5	17-10	9-1	11-6	14-1	16-3
	Douglas Fir-Larch	#3	7-6	9-6	11-8	13-6	6-10	8-8	10-7	12-4
	Hem-Fir	SS	9-9	12-10	16-5	19-11	9-9	12-10	16-5	19-11
	Hem-Fir	#1	9-6	12-7	16-0	18-7	9-6	12-0	14-8	17-0
	Hem-Fir	#2	9-1	12-0	15-2	17-7	8-11	11-4	13-10	16-1
	Hem-Fir	#3	7-6	9-6	11-8	13-6	6-10	8-8	10-7	12-4
	Southern Pine	SS	10-2	13-4	17-0	20-9	10-2	13-4	17-0	20-9
	Southern Pine	#1	9-11	13-1	16-9	20-4	9-11	13-1	16-4	19-6
	Southern Pine	#2	9-9	12-10	16-1	18-10	9-6	12-4	14-8	17-2
	Southern Pine	#3	8-1	10-3	12-2	14-6	7-4	9-5	11-1	13-2
	Spruce-Pine-Fir	SS	9-6	12-7	16-0	19-6	9-6	12-7	16-0	19-6
	Spruce-Pine-Fir	#1	9-4	12-3	15-5	17-10	9-1	11-6	14-1	16-3
	Spruce-Pine-Fir	#2	9-4	12-3	15-5	17-10	9-1	11-6	14-1	16-3
	Spruce-Pine-Fir	#3	7-6	9-6	11-8	13-6	6-10	8-8	10-7	12-4

(continued)

TABLE 2308.8(2)—continued
FLOOR JOIST SPANS FOR COMMON LUMBER SPECIES
(Residential Living Areas, Live Load = 40 psf, L/Δ = 360)

JOIST SPACING (inches)	SPECIES AND GRADE		DEAD LOAD = 10 psf				DEAD LOAD = 20 psf			
			2 x 6	2 x 8	2 x 10	2 x 12	2 x 6	2 x 8	2 x 10	2 x 12
			Maximum floor joist spans							
			(ft. - in.)	(ft. - in.)	(ft. - in.)	(ft. - in.)	(ft. - in.)	(ft. - in.)	(ft. - in.)	(ft. - in.)
19.2	Douglas Fir-Larch	SS	9-8	12-10	16-4	19-10	9-8	12-10	16-4	19-2
	Douglas Fir-Larch	#1	9-4	12-4	15-0	17-5	8-10	11-3	13-8	15-11
	Douglas Fir-Larch	#2	9-1	11-6	14-1	16-3	8-3	10-6	12-10	14-10
	Douglas Fir-Larch	#3	6-10	8-8	10-7	12-4	6-3	7-11	9-8	11-3
	Hem-Fir	SS	9-2	12-1	15-5	18-9	9-2	12-1	15-5	18-9
	Hem-Fir	#1	9-0	11-10	14-8	17-0	8-8	10-11	13-4	15-6
	Hem-Fir	#2	8-7	11-3	13-10	16-1	8-2	10-4	12-8	14-8
	Hem-Fir	#3	6-10	8-8	10-7	12-4	6-3	7-11	9-8	11-3
	Southern Pine	SS	9-6	12-7	16-0	19-6	9-6	12-7	16-0	19-6
	Southern Pine	#1	9-4	12-4	15-9	19-2	9-4	12-4	14-11	17-9
	Southern Pine	#2	9-2	12-1	14-8	17-2	8-8	11-3	13-5	15-8
	Southern Pine	#3	7-4	9-5	11-1	13-2	6-9	8-7	10-1	12-1
	Spruce-Pine-Fir	SS	9-0	11-10	15-1	18-4	9-0	11-10	15-1	17-9
	Spruce-Pine-Fir	#1	8-9	11-6	14-1	16-3	8-3	10-6	12-10	14-10
	Spruce-Pine-Fir	#2	8-9	11-6	14-1	16-3	8-3	10-6	12-10	14-10
	Spruce-Pine-Fir	#3	6-10	8-8	10-7	12-4	6-3	7-11	9-8	11-3
24	Douglas Fir-Larch	SS	9-0	11-11	15-2	18-5	9-0	11-11	14-9	17-1
	Douglas Fir-Larch	#1	8-8	11-0	13-5	15-7	7-11	10-0	12-3	14-3
	Douglas Fir-Larch	#2	8-1	10-3	12-7	14-7	7-5	9-5	11-6	13-4
	Douglas Fir-Larch	#3	6-2	7-9	9-6	11-0	5-7	7-1	8-8	10-1
	Hem-Fir	SS	8-6	11-3	14-4	17-5	8-6	11-3	14-4	16-10[a]
	Hem-Fir	#1	8-4	10-9	13-1	15-2	7-9	9-9	11-11	13-10
	Hem-Fir	#2	7-11	10-2	12-5	14-4	7-4	9-3	11-4	13-1
	Hem-Fir	#3	6-2	7-9	9-6	11-0	5-7	7-1	8-8	10-1
	Southern Pine	SS	8-10	11-8	14-11	18-1	8-10	11-8	14-11	18-1
	Southern Pine	#1	8-8	11-5	14-7	17-5	8-8	11-3	13-4	15-11
	Southern Pine	#2	8-6	11-0	13-1	15-5	7-9	10-0	12-0	14-0
	Southern Pine	#3	6-7	8-5	9-11	11-10	6-0	7-8	9-1	10-9
	Spruce-Pine-Fir	SS	8-4	11-0	14-0	17-0	8-4	11-0	13-8	15-11
	Spruce-Pine-Fir	#1	8-1	10-3	12-7	14-7	7-5	9-5	11-6	13-4
	Spruce-Pine-Fir	#2	8-1	10-3	12-7	14-7	7-5	9-5	11-6	13-4
	Spruce-Pine-Fir	#3	6-2	7-9	9-6	11-0	5-7	7-1	8-8	10-1

Check sources for availability of lumber in lengths greater than 20 feet.

For SI: 1 inch = 25.4 mm, 1 foot = 304.8 mm, 1 pound per square foot = 47.8 N/m².

a. End bearing length shall be increased to 2 inches.

TABLE 2308.9.1
SIZE, HEIGHT AND SPACING OF WOOD STUDS

STUD SIZE (inches)	BEARING WALLS				NONBEARING WALLS	
	Laterally unsupported stud height[a] (feet)	Supporting roof and ceiling only	Supporting one floor, roof and ceiling	Supporting two floors, roof and ceiling	Laterally unsupported stud height[a] (feet)	Spacing (inches)
		Spacing (inches)				
2 × 3[b]	—	—	—	—	10	16
2 × 4	10	24	16	—	14	24
3 × 4	10	24	24	16	14	24
2 × 5	10	24	24	—	16	24
2 × 6	10	24	24	16	20	24

For SI: 1 inch = 25.4 mm, 1 foot = 304.8 mm.

a. Listed heights are distances between points of lateral support placed perpendicular to the plane of the wall. Increases in unsupported height are permitted where justified by an analysis.

b. Shall not be used in exterior walls.

TABLE 2308.9.3(1)
BRACED WALL PANELS[a]

SEISMIC DESIGN CATEGORY	CONDITION	CONSTRUCTION METHODS[b,c]								BRACED PANEL LOCATION AND LENGTH[d]
		1	2	3	4	5	6	7	8	
A and B	One story, top of two or three story	X	X	X	X	X	X	X	X	Located in accordance with Section 2308.9.3 and not more than 25 feet on center.
	First story of two story or second story of three story	X	X	X	X	X	X	X	X	
	First story of three story	—	X	X	X	X[e]	X	X	X	
C	One story or top of two story	—	X	X	X	X	X	X	X	Located in accordance with Section 2308.9.3 and not more than 25 feet on center.
	First story of two story	—	X	X	X	X[e]	X	X	X	Located in accordance with Section 2308.9.3 and not more than 25 feet on center, but total length shall not be less than 25% of building length[f].

For SI: 1 inch = 25.4 mm, 1 foot = 304.8 mm.

a. This table specifies minimum requirements for braced panels that form interior or exterior braced wall lines.

b. See Section 2308.9.3 for full description.

c. See Sections 2308.9.3.1 and 2308.9.3.2 for alternative braced panel requirements.

d. Building length is the dimension parallel to the braced wall length.

e. Gypsum wallboard applied to framing supports that are spaced at 16 inches on center.

f. The required lengths shall be doubled for gypsum board applied to only one face of a braced wall panel.

TABLE 2308.9.3(2)
EXPOSED PLYWOOD PANEL SIDING

MINIMUM THICKNESS[a] (inch)	MINIMUM NUMBER OF PLIES	STUD SPACING (inches) Plywood siding applied directly to studs or over sheathing
$3/_8$	3	16[b]
$1/_2$	4	24

For SI: 1 inch = 25.4 mm.

a. Thickness of grooved panels is measured at bottom of grooves.

b. Spans are permitted to be 24 inches if plywood siding applied with face grain perpendicular to studs or over one of the following: (1) 1-inch board sheathing, (2) $7/_{16}$ -inch wood structural panel sheathing or (3) $3/_8$-inch wood structural panel sheathing with strength axis (which is the long direction of the panel unless otherwise marked) of sheathing perpendicular to studs.

TABLE 2308.9.3(3)
WOOD STRUCTURAL PANEL WALL SHEATHING[b]
(Not Exposed to the Weather, Strength Axis Parallel or Perpendicular to Studs Except as Indicated Below)

MINIMUM THICKNESS (inch)	PANEL SPAN RATING	STUD SPACING (inches)		
		Siding nailed to studs	Nailable sheathing	
			Sheathing parallel to studs	Sheathing perpendicular to studs
$5/16$	12/0, 16/0, 20/0 Wall–16" o.c.	16	—	16
$3/8$, $15/32$, $1/2$	16/0, 20/0, 24/0, 32/16 Wall–24" o.c.	24	16	24
$7/16$, $15/32$, $1/2$	24/0, 24/16, 32/16 Wall–24" o.c.	24	24[a]	24

For SI: 1 inch = 25.4 mm.
a. Plywood shall consist of four or more plies.
b. Blocking of horizontal joints shall not be required except as specified in Sections 2306.4 and 2308.12.4.

TABLE 2308.9.3(4)
ALLOWABLE SPANS FOR PARTICLEBOARD WALL SHEATHING
(Not Exposed to the Weather, Long Dimension of the Panel Parallel or Perpendicular to Studs)

GRADE	THICKNESS (inch)	STUD SPACING (inches)	
		Siding nailed to studs	Sheathing under coverings specified in Section 2308.9.3 parallel or perpendicular to studs
M-S "Exterior Glue" and M-2 "Exterior Glue"	$3/8$	16	—
	$1/2$	16	16

For SI: 1 inch = 25.4 mm.

TABLE 2308.9.3(5)
HARDBOARD SIDING

SIDING	MINIMUM NOMINAL THICKNESS (inch)	2 × 4 FRAMING MAXIMUM SPACING	NAIL SIZE[a,b,d]	NAIL SPACING	
				General	Bracing panels[c]
1. Lap siding					
Direct to studs	³/₈	16″ o.c.	8d	16″ o.c.	Not applicable
Over sheathing	³/₈	16″ o.c.	10d	16″ o.c.	Not applicable
2. Square edge panel siding					
Direct to studs	³/₈	24″ o.c.	6d	6″ o.c. edges; 12″ o.c. at intermediate supports	4″ o.c. edges; 8″ o.c. at intermediate supports
Over sheathing	³/₈	24″ o.c.	8d	6″ o.c. edges; 12″ o.c. at intermediate supports	4″ o.c. edges; 8″ o.c. at intermediate supports
3. Shiplap edge panel siding					
Direct to studs	³/₈	16″ o.c.	6d	6″ o.c. edges; 12″ o.c. at intermediate supports	4″ o.c. edges; 8″ o.c. at intermediate supports
Over sheathing	³/₈	16″ o.c.	8d	6″ o.c. edges; 12″ o.c. At intermediate supports	4″ o.c. edges; 8″ o.c. at intermediate supports

For SI: 1 inch = 25.4 mm.
a. Nails shall be corrosion resistant.
b. Minimum acceptable nail dimensions:

	Panel Siding (inch)	Lap Siding (inch)
Shank diameter	0.092	0.099
Head diameter	0.225	0.240

c. Where used to comply with Section 2308.9.3.
d. Nail length must accommodate the sheathing and penetrate framing $1^1/_2$ inches.

TABLE 2308.9.5
HEADER AND GIRDER SPANS[a] FOR EXTERIOR BEARING WALLS
(Maximum Spans for Douglas Fir-Larch, Hem-Fir, Southern Pine and Spruce-Pine-Fir[b] and Required Number of Jack Studs)

HEADERS SUPPORTING	SIZE	GROUND SNOW LOAD (psf)[c]											
		30						50					
		Building width[e] (feet)											
		20		28		36		20		28		36	
		Span	NJ[d]	Span	NJ[d]	Span	NJ[d]	Span	NJ[d]	Span	NJ[d]	Span	NJ[d]
Roof & Ceiling	2-2×4	3-6	1	3-2	1	2-10	1	3-2	1	2-9	1	2-6	1
	2-2×6	5-5	1	4-8	1	4-2	1	4-8	1	4-1	1	3-8	2
	2-2×8	6-10	1	5-11	2	5-4	2	5-11	2	5-2	2	4-7	2
	2-2×10	8-5	2	7-3	2	6-6	2	7-3	2	6-3	2	5-7	2
	2-2×12	9-9	2	8-5	2	7-6	2	8-5	2	7-3	2	6-6	2
	3-2×8	8-4	1	7-5	1	6-8	1	7-5	1	6-5	2	5-9	2
	3-2×10	10-6	1	9-1	2	8-2	2	9-1	2	7-10	2	7-0	2
	3-2×12	12-2	2	10-7	2	9-5	2	10-7	2	9-2	2	8-2	2
	4-2×8	9-2	1	8-4	1	7-8	1	8-4	1	7-5	1	6-8	1
	4-2×10	11-8	1	10-6	1	9-5	1	10-6	1	9-1	2	8-2	2
	4-2×12	14-1	1	12-2	2	10-11	2	12-2	2	10-7	2	9-5	2
Roof, Ceiling & 1 Center-Bearing Floor	2-2×4	3-1	1	2-9	1	2-5	1	2-9	1	2-5	1	2-2	1
	2-2×6	4-6	1	4-0	1	3-7	1	4-1	1	3-7	1	3-3	2
	2-2×8	5-9	2	5-0	2	4-6	2	5-2	2	4-6	2	4-1	2
	2-2×10	7-0	2	6-2	2	5-6	2	6-4	2	5-6	2	5-0	2
	2-2×12	8-1	2	7-1	2	6-5	2	7-4	2	6-5	2	5-9	3
	3-2×8	7-2	1	6-3	1	5-8	2	6-5	2	5-8	2	5-1	2
	3-2×10	8-9	2	7-8	2	6-11	2	7-11	2	6-11	2	6-3	2
	3-2×12	10-2	2	8-11	2	8-0	2	9-2	2	8-0	2	7-3	2
	4-2×8	8-1	1	7-3	1	6-7	1	7-5	1	6-6	1	5-11	1
	4-2×10	10-1	1	8-10	1	8-0	2	9-1	2	8-0	2	7-2	2
	4-2×12	11-9	2	10-3	2	9-3	2	10-7	2	9-3	2	8-4	2
Roof, Ceiling & 1 Clear Span Floor	2-2×4	2-8	1	2-4	1	2-1	1	2-7	1	2-3	1	2-0	1
	2-2×6	3-11	1	3-5	2	3-0	2	3-10	2	3-4	2	3-0	2
	2-2×8	5-0	2	4-4	2	3-10	2	4-10	2	4-2	2	3-9	2
	2-2×10	6-1	2	5-3	2	4-8	2	5-11	2	5-1	2	4-7	3
	2-2×12	7-1	2	6-1	3	5-5	3	6-10	2	5-11	3	5-4	3
	3-2×8	6-3	2	5-5	2	4-10	2	6-1	2	5-3	2	4-8	2
	3-2×10	7-7	2	6-7	2	5-11	2	7-5	2	6-5	2	5-9	2
	3-2×12	8-10	2	7-8	2	6-10	2	8-7	2	7-5	2	6-8	2
	4-2×8	7-2	1	6-3	2	5-7	2	7-0	1	6-1	1	5-5	2
	4-2×10	8-9	2	7-7	2	6-10	2	8-7	2	7-5	2	6-7	2
	4-2×12	10-2	2	8-10	2	7-11	2	9-11	2	8-7	2	7-8	2

(continued)

TABLE 2308.9.5—continued
HEADER AND GIRDER SPANS^a FOR EXTERIOR BEARING WALLS
(Maximum Spans for Douglas Fir-Larch, Hem-Fir, Southern Pine and Spruce-Pine-Fir^b and Required Number of Jack Studs)

HEADERS SUPPORTING	SIZE	GROUND SNOW LOAD (psf)^e											
		30						50					
		Building width^c (feet)											
		20		28		36		20		28		36	
		Span	NJ^d	Span	NJ^d	Span	NJ^d	Span	NJ^d	Span	NJ^d	Span	NJ^d
Roof, Ceiling & 2 Center-Bearing Floors	2-2 × 4	2-7	1	2-3	1	2-0	1	2-6	1	2-2	1	1-11	1
	2-2 × 6	3-9	2	3-3	2	2-11	2	3-8	2	3-2	2	2-10	2
	2-2 × 8	4-9	2	4-2	2	3-9	2	4-7	2	4-0	2	3-8	2
	2-2 × 10	5-9	2	5-1	2	4-7	3	5-8	2	4-11	2	4-5	3
	2-2 × 12	6-8	2	5-10	3	5-3	3	6-6	2	5-9	3	5-2	3
	3-2 × 8	5-11	2	5-2	2	4-8	2	5-9	2	5-1	2	4-7	2
	3-2 × 10	7-3	2	6-4	2	5-8	2	7-1	2	6-2	2	5-7	2
	3-2 × 12	8-5	2	7-4	2	6-7	2	8-2	2	7-2	2	6-5	3
	4-2 × 8	6-10	1	6-0	2	5-5	2	6-8	1	5-10	2	5-3	2
	4-2 × 10	8-4	2	7-4	2	6-7	2	8-2	2	7-2	2	6-5	2
	4-2 × 12	9-8	2	8-6	2	7-8	2	9-5	2	8-3	2	7-5	2
Roof, Ceiling & 2 Clear Span Floors	2-2 × 4	2-1	1	1-8	2	1-6	2	2-0	1	1-8	1	1-5	2
	2-2 × 6	3-1	2	2-8	2	2-4	2	3-0	2	2-7	2	2-3	2
	2-2 × 8	3-10	2	3-4	2	3-0	3	3-10	2	3-4	2	2-11	3
	2-2 × 10	4-9	2	4-1	3	3-8	3	4-8	2	4-0	3	3-7	3
	2-2 × 12	5-6	3	4-9	3	4-3	3	5-5	3	4-8	3	4-2	3
	3-2 × 8	4-10	2	4-2	2	3-9	2	4-9	2	4-1	2	3-8	2
	3-2 × 10	5-11	2	5-1	2	4-7	3	5-10	2	5-0	2	4-6	3
	3-2 × 12	6-10	2	5-11	3	5-4	3	6-9	2	5-10	3	5-3	3
	4-2 × 8	5-7	2	4-10	2	4-4	2	5-6	2	4-9	2	4-3	2
	4-2 × 10	6-10	2	5-11	2	5-3	2	6-9	2	5-10	2	5-2	2
	4-2 × 12	7-11	2	6-10	2	6-2	2	7-9	2	6-9	2	6-0	3

For SI: 1 inch = 25.4 mm, 1 foot = 304.8 mm, 1 pound per square foot = 47.8 N/m².

a. Spans are given in feet and inches (ft-in).

b. Tabulated values are for No. 2 grade lumber.

c. Building width is measured perpendicular to the ridge. For widths between those shown, spans are permitted to be interpolated.

d. NJ - Number of jack studs required to support each end. Where the number of required jack studs equals one, the header is permitted to be supported by an approved framing anchor attached to the full-height wall stud and to the header.

e. Use 30 pounds per square foot ground snow load for cases in which ground snow load is less than 30 pounds per square foot and the roof live load is equal to or less than 20 pounds per square foot.

TABLE 2308.9.6
HEADER AND GIRDER SPANS[a] FOR INTERIOR BEARING WALLS
(Maximum Spans for Douglas Fir-Larch, Hem-Fir, Southern Pine and Spruce-Pine-Fir[b] and Required Number of Jack Studs)

HEADERS AND GIRDERS SUPPORTING	SIZE	BUILDING WIDTH[c] (feet)					
		20		28		36	
		Span	NJ[d]	Span	NJ[d]	Span	NJ[d]
One Floor Only	2-2 × 4	3-1	1	2-8	1	2-5	1
	2-2 × 6	4-6	1	3-11	1	3-6	1
	2-2 × 8	5-9	1	5-0	2	4-5	2
	2-2 ×10	7-0	2	6-1	2	5-5	2
	2-2 ×12	8-1	2	7-0	2	6-3	2
	3-2 × 8	7-2	1	6-3	1	5-7	2
	3-2 ×10	8-9	1	7-7	2	6-9	2
	3-2 ×12	10-2	2	8-10	2	7-10	2
	4-2 × 8	9-0	1	7-8	1	6-9	1
	4-2 ×10	10-1	1	8-9	1	7-10	2
	4-2 ×12	11-9	1	10-2	2	9-1	2
Two Floors	2-2 × 4	2-2	1	1-10	1	1-7	1
	2-2 × 6	3-2	2	2-9	2	2-5	2
	2-2 × 8	4-1	2	3-6	2	3-2	2
	2-2 ×10	4-11	2	4-3	2	3-10	3
	2-2 ×12	5-9	2	5-0	3	4-5	3
	3-2 × 8	5-1	2	4-5	2	3-11	2
	3-2 ×10	6-2	2	5-4	2	4-10	2
	3-2 ×12	7-2	2	6-3	2	5-7	3
	4-2 × 8	6-1	1	5-3	2	4-8	2
	4-2 ×10	7-2	2	6-2	2	5-6	2
	4-2 ×12	8-4	2	7-2	2	6-5	2

For SI: 1 inch = 25.4 mm, 1 foot = 304.8 mm.
a. Spans are given in feet and inches (ft-in).
b. Tabulated values are for No. 2 grade lumber.
c. Building width is measured perpendicular to the ridge. For widths between those shown, spans are permitted to be interpolated.
d. NJ - Number of jack studs required to support each end. Where the number of required jack studs equals one, the headers are permitted to be supported by an approved framing anchor attached to the full-height wall stud and to the header.

TABLE 2308.10.2(1)
CEILING JOIST SPANS FOR COMMON LUMBER SPECIES
(Uninhabitable Attics Without Storage, Live Load = 10 pounds psf, L/Δ = 240)

CEILING JOIST SPACING (inches)	SPECIES AND GRADE		DEAD LOAD = 5 pounds per square foot			
			Maximum ceiling joist spans			
			2 × 4	2 × 6	2 × 8	2 × 10
			(ft. - in.)	(ft. - in.)	(ft. - in.)	(ft. - in.)
12	Douglas Fir-Larch	SS	13-2	20-8	Note a	Note a
	Douglas Fir-Larch	#1	12-8	19-11	Note a	Note a
	Douglas Fir-Larch	#2	12-5	19-6	25-8	Note a
	Douglas Fir-Larch	#3	10-10	15-10	20-1	24-6
	Hem-Fir	SS	12-5	19-6	25-8	Note a
	Hem-Fir	#1	12-2	19-1	25-2	Note a
	Hem-Fir	#2	11-7	18-2	24-0	Note a
	Hem-Fir	#3	10-10	15-10	20-1	24-6
	Southern Pine	SS	12-11	20-3	Note a	Note a
	Southern Pine	#1	12-8	19-11	Note a	Note a
	Southern Pine	#2	12-5	19-6	25-8	Note a
	Southern Pine	#3	11-6	17-0	21-8	25-7
	Spruce-Pine-Fir	SS	12-2	19-1	25-2	Note a
	Spruce-Pine-Fir	#1	11-10	18-8	24-7	Note a
	Spruce-Pine-Fir	#2	11-10	18-8	24-7	Note a
	Spruce-Pine-Fir	#3	10-10	15-10	20-1	24-6
16	Douglas Fir-Larch	SS	11-11	18-9	24-8	Note a
	Douglas Fir-Larch	#1	11-6	18-1	23-10	Note a
	Douglas Fir-Larch	#2	11-3	17-8	23-0	Note a
	Douglas Fir-Larch	#3	9-5	13-9	17-5	21-3
	Hem-Fir	SS	11-3	17-8	23-4	Note a
	Hem-Fir	#1	11-0	17-4	22-10	Note a
	Hem-Fir	#2	10-6	16-6	21-9	Note a
	Hem-Fir	#3	9-5	13-9	17-5	21-3
	Southern Pine	SS	11-9	18-5	24-3	Note a
	Southern Pine	#1	11-6	18-1	23-1	Note a
	Southern Pine	#2	11-3	17-8	23-4	Note a
	Southern Pine	#3	10-0	14-9	18-9	22-2
	Spruce-Pine-Fir	SS	11-0	17-4	22-10	Note a
	Spruce-Pine-Fir	#1	10-9	16-11	22-4	Note a
	Spruce-Pine-Fir	#2	10-9	16-11	22-4	Note a
	Spruce-Pine-Fir	#3	9-5	13-9	17-5	21-3

(continued)

TABLE 2308.10.2(1)—continued
CEILING JOIST SPANS FOR COMMON LUMBER SPECIES
(Uninhabitable Attics Without Storage, Live Load = 10 pounds psf, L/Δ = 240)

CEILING JOIST SPACING (inches)	SPECIES AND GRADE		DEAD LOAD = 5 pounds per square foot			
			Maximum ceiling joist spans			
			2 × 4	2 × 6	2 × 8	2 × 10
			(ft. - in.)	(ft. - in.)	(ft. - in.)	(ft. - in.)
19.2	Douglas Fir-Larch	SS	11-3	17-8	23-3	Note a
	Douglas Fir-Larch	#1	10-10	17-0	22-5	Note a
	Douglas Fir-Larch	#2	10-7	16-7	21-0	25-8
	Douglas Fir-Larch	#3	8-7	12-6	15-10	19-5
	Hem-Fir	SS	10-7	16-8	21-11	Note a
	Hem-Fir	#1	10-4	16-4	21-6	Note a
	Hem-Fir	#2	9-11	15-7	20-6	25-3
	Hem-Fir	#3	8-7	12-6	15-10	19-5
	Southern Pine	SS	11-0	17-4	22-10	Note a
	Southern Pine	#1	10-10	17-0	22-5	Note a
	Southern Pine	#2	10-7	16-8	21-11	Note a
	Southern Pine	#3	9-1	13-6	17-2	20-3
	Spruce-Pine-Fir	SS	10-4	16-4	21-6	Note a
	Spruce-Pine-Fir	#1	10-2	15-11	21-0	25-8
	Spruce-Pine-Fir	#2	10-2	15-11	21-0	25-8
	Spruce-Pine-Fir	#3	8-7	12-6	15-10	19-5
24	Douglas Fir-Larch	SS	10-5	16-4	21-7	Note a
	Douglas Fir-Larch	#1	10-0	15-9	20-1	24-6
	Douglas Fir-Larch	#2	9-10	14-10	18-9	22-11
	Douglas Fir-Larch	#3	7-8	11-2	14-2	17-4
	Hem-Fir	SS	9-10	15-6	20-5	Note a
	Hem-Fir	#1	9-8	15-2	19-7	23-11
	Hem-Fir	#2	9-2	14-5	18-6	22-7
	Hem-Fir	#3	7-8	11-2	14-2	17-4
	Southern Pine	SS	10-3	16-1	21-2	Note a
	Southern Pine	#1	10-0	15-9	20-10	Note a
	Southern Pine	#2	9-10	15-6	20-1	23-11
	Southern Pine	#3	8-2	12-0	15-4	18-1
	Spruce-Pine-Fir	SS	9-8	15-2	19-11	25-5
	Spruce-Pine-Fir	#1	9-5	14-9	18-9	22-11
	Spruce-Pine-Fir	#2	9-5	14-9	18-9	22-11
	Spruce-Pine-Fir	#3	7-8	11-2	14-2	17-4

For SI: 1 inch = 25.4 mm, 1 foot = 304.8 mm, 1 pound per square foot = 47.8 N/m².
a. Span exceeds 26 feet in length. Check sources for availability of lumber in lengths greater than 20 feet.

TABLE 2308.10.2(2)
CEILING JOIST SPANS FOR COMMON LUMBER SPECIES
(Uninhabitable Attics With Limited Storage, Live Load = 20 pounds per square foot, $L/\Delta = 240$)

CEILING JOIST SPACING (inches)	SPECIES AND GRADE		DEAD LOAD = 10 pounds per square foot			
			Maximum ceiling joist spans			
			2 × 4	2 × 6	2 × 8	2 × 10
			(ft. - in.)	(ft. - in.)	(ft. - in.)	(ft. - in.)
12	Douglas Fir-Larch	SS	10-5	16-4	21-7	Note a
	Douglas Fir-Larch	#1	10-0	15-9	20-1	24-6
	Douglas Fir-Larch	#2	9-10	14-10	18-9	22-11
	Douglas Fir-Larch	#3	7-8	11-2	14-2	17-4
	Hem-Fir	SS	9-10	15-6	20-5	Note a
	Hem-Fir	#1	9-8	15-2	19-7	23-11
	Hem-Fir	#2	9-2	14-5	18-6	22-7
	Hem-Fir	#3	7-8	11-2	14-2	17-4
	Southern Pine	SS	10-3	16-1	21-2	Note a
	Southern Pine	#1	10-0	15-9	20-10	Note a
	Southern Pine	#2	9-10	15-6	20-1	23-11
	Southern Pine	#3	8-2	12-0	15-4	18-1
	Spruce-Pine-Fir	SS	9-8	15-2	19-11	25-5
	Spruce-Pine-Fir	#1	9-5	14-9	18-9	22-11
	Spruce-Pine-Fir	#2	9-5	14-9	18-9	22-11
	Spruce-Pine-Fir	#3	7-8	11-2	14-2	17-4
16	Douglas Fir-Larch	SS	9-6	14-11	19-7	25-0
	Douglas Fir-Larch	#1	9-1	13-9	17-5	21-3
	Douglas Fir-Larch	#2	8-9	12-10	16-3	19-10
	Douglas Fir-Larch	#3	6-8	9-8	12-4	15-0
	Hem-Fir	SS	8-11	14-1	18-6	23-8
	Hem-Fir	#1	8-9	13-5	16-10	20-8
	Hem-Fir	#2	8-4	12-8	16-0	19-7
	Hem-Fir	#3	6-8	9-8	12-4	15-0
	Southern Pine	SS	9-4	14-7	19-3	24-7
	Southern Pine	#1	9-1	14-4	18-11	23-1
	Southern Pine	#2	8-11	13-6	17-5	20-9
	Southern Pine	#3	7-1	10-5	13-3	15-8
	Spruce-Pine-Fir	SS	8-9	13-9	18-1	23-1
	Spruce-Pine-Fir	#1	8-7	12-10	16-3	19-10
	Spruce-Pine-Fir	#2	8-7	12-10	16-3	19-10
	Spruce-Pine-Fir	#3	6-8	9-8	12-4	15-0

(continued)

TABLE 2308.10.2(2)—continued
CEILING JOIST SPANS FOR COMMON LUMBER SPECIES
(Uninhabitable Attics With Limited Storage, Live Load = 20 pounds per square foot, $L/\Delta = 240$)

CEILING JOIST SPACING (inches)	SPECIES AND GRADE		DEAD LOAD = 10 pounds per square foot			
			2 × 4	2 × 6	2 × 8	2 × 10
			Maximum ceiling joist spans			
			(ft. - in.)	(ft. - in.)	(ft. - in.)	(ft. - in.)
19.2	Douglas Fir-Larch	SS	8-11	14-0	18-5	23-4
	Douglas Fir-Larch	#1	8-7	12-6	15-10	19-5
	Douglas Fir-Larch	#2	8-0	11-9	14-10	18-2
	Douglas Fir-Larch	#3	6-1	8-10	11-3	13-8
	Hem-Fir	SS	8-5	13-3	17-5	22-3
	Hem-Fir	#1	8-3	12-3	15-6	18-11
	Hem-Fir	#2	7-10	11-7	14-8	17-10
	Hem-Fir	#3	6-1	8-10	11-3	13-8
	Southern Pine	SS	8-9	13-9	18-1	23-1
	Southern Pine	#1	8-7	13-6	17-9	21-1
	Southern Pine	#2	8-5	12-3	15-10	18-11
	Southern Pine	#3	6-5	9-6	12-1	14-4
	Spruce-Pine-Fir	SS	8-3	12-11	17-1	21-8
	Spruce-Pine-Fir	#1	8-0	11-9	14-10	18-2
	Spruce-Pine-Fir	#2	8-0	11-9	14-10	18-2
	Spruce-Pine-Fir	#3	6-1	8-10	11-3	13-8
24	Douglas Fir-Larch	SS	8-3	13-0	17-1	20-11
	Douglas Fir-Larch	#1	7-8	11-2	14-2	17-4
	Douglas Fir-Larch	#2	7-2	10-6	13-3	16-3
	Douglas Fir-Larch	#3	5-5	7-11	10-0	12-3
	Hem-Fir	SS	7-10	12-3	16-2	20-6
	Hem-Fir	#1	7-6	10-11	13-10	16-11
	Hem-Fir	#2	7-1	10-4	13-1	16-0
	Hem-Fir	#3	5-5	7-11	10-0	12-3
	Southern Pine	SS	8-1	12-9	16-10	21-6
	Southern Pine	#1	8-0	12-6	15-10	18-10
	Southern Pine	#2	7-8	11-0	14-2	16-11
	Southern Pine	#3	5-9	8-6	10-10	12-10
	Spruce-Pine-Fir	SS	7-8	12-0	15-10	19-5
	Spruce-Pine-Fir	#1	7-2	10-6	13-3	16-3
	Spruce-Pine-Fir	#2	7-2	10-6	13-3	16-3
	Spruce-Pine-Fir	#3	5-5	7-11	10-0	12-3

For SI: 1 inch = 25.4 mm, 1 foot = 304.8 mm, 1 pound per square foot = 47.8 N/m².
a. Span exceeds 26 feet in length. Check sources for availability of lumber in lengths greater than 20 feet.

TABLE 2308.10.3(1)
RAFTER SPANS FOR COMMON LUMBER SPECIES
(Roof Live Load = 20 pounds per square foot, Ceiling Not Attached to Rafters, $L/\Delta = 180$)

RAFTER SPACING (inches)	SPECIES AND GRADE		DEAD LOAD = 10 pounds per square foot					DEAD LOAD = 20 pounds per square foot				
			Maximum rafter spans									
			2 × 4	2 × 6	2 × 8	2 × 10	2 × 12	2 × 4	2 × 6	2 × 8	2 × 10	2 × 12
			(ft. - in.)	(ft. - in.)	(ft. - in.)	(ft. - in.)	(ft. - in.)	(ft. - in.)	(ft. - in.)	(ft. - in.)	(ft. - in.)	(ft. - in.)
12	Douglas Fir-Larch	SS	11-6	18-0	23-9	Note a	Note a	11-6	18-0	23-5	Note a	Note a
	Douglas Fir-Larch	#1	11-1	17-4	22-5	Note a	Note a	10-6	15-4	19-5	23-9	Note a
	Douglas Fir-Larch	#2	10-10	16-7	21-0	25-8	Note a	9-10	14-4	18-2	22-3	25-9
	Douglas Fir-Larch	#3	8-7	12-6	15-10	19-5	22-6	7-5	10-10	13-9	16-9	19-6
	Hem-Fir	SS	10-10	17-0	22-5	Note a	Note a	10-10	17-0	22-5	Note a	Note a
	Hem-Fir	#1	10-7	16-8	21-10	Note a	Note a	10-3	14-11	18-11	23-2	Note a
	Hem-Fir	#2	10-1	15-11	20-8	25-3	Note a	9-8	14-2	17-11	21-11	25-5
	Hem-Fir	#3	8-7	12-6	15-10	19-5	22-6	7-5	10-10	13-9	16-9	19-6
	Southern Pine	SS	11-3	17-8	23-4	Note a	Note a	11-3	17-8	23-4	Note a	Note a
	Southern Pine	#1	11-1	17-4	22-11	Note a	Note a	11-1	17-3	21-9	25-10	Note a
	Southern Pine	#2	10-10	17-0	22-5	Note a	Note a	10-6	15-1	19-5	23-2	Note a
	Southern Pine	#3	9-1	13-6	17-2	20-3	24-1	7-11	11-8	14-10	17-6	20-11
	Spruce-Pine-Fir	SS	10-7	16-8	21-11	Note a	Note a	10-7	16-8	21-9	Note a	Note a
	Spruce-Pine-Fir	#1	10-4	16-3	21-0	25-8	Note a	9-10	14-4	18-2	22-3	25-9
	Spruce-Pine-Fir	#2	10-4	16-3	21-0	25-8	Note a	9-10	14-4	18-2	22-3	25-9
	Spruce-Pine-Fir	#3	8-7	12-6	15-10	19-5	22-6	7-5	10-10	13-9	16-9	19-6
16	Douglas Fir-Larch	SS	10-5	16-4	21-7	Note a	Note a	10-5	16-0	20-3	24-9	Note a
	Douglas Fir-Larch	#1	10-0	15-4	19-5	23-9	Note a	9-1	13-3	16-10	20-7	23-10
	Douglas Fir-Larch	#2	9-10	14-4	18-2	22-3	25-9	8-6	12-5	15-9	19-3	22-4
	Douglas Fir-Larch	#3	7-5	10-10	13-9	16-9	19-6	6-5	9-5	11-11	14-6	16-10
	Hem-Fir	SS	9-10	15-6	20-5	Note a	Note a	9-10	15-6	19-11	24-4	Note a
	Hem-Fir	#1	9-8	14-11	18-11	23-2	Note a	8-10	12-11	16-5	20-0	23-3
	Hem-Fir	#2	9-2	14-2	17-11	21-11	25-5	8-5	12-3	15-6	18-11	22-0
	Hem-Fir	#3	7-5	10-10	13-9	16-9	19-6	6-5	9-5	11-11	14-6	16-10
	Southern Pine	SS	10-3	16-1	21-2	Note a	Note a	10-3	16-1	21-2	Note a	Note a
	Southern Pine	#1	10-0	15-9	20-10	25-10	Note a	10-0	15-0	18-10	22-4	Note a
	Southern Pine	#2	9-10	15-1	19-5	23-2	Note a	9-1	13-0	16-10	20-1	23-7
	Southern Pine	#3	7-11	11-8	14-10	17-6	20-11	6-10	10-1	12-10	15-2	18-1
	Spruce-Pine-Fir	SS	9-8	15-2	19-11	25-5	Note a	9-8	14-10	18-10	23-0	Note a
	Spruce-Pine-Fir	#1	9-5	14-4	18-2	22-3	25-9	8-6	12-5	15-9	19-3	22-4
	Spruce-Pine-Fir	#2	9-5	14-4	18-2	22-3	25-9	8-6	12-5	15-9	19-3	22-4
	Spruce-Pine-Fir	#3	7-5	10-10	13-9	16-9	19-6	6-5	9-5	11-11	14-6	16-10

(continued)

TABLE 2308.10.3(1)—continued
RAFTER SPANS FOR COMMON LUMBER SPECIES
(Roof Live Load = 20 pounds per square foot, Ceiling Not Attached to Rafters, L/Δ = 180)

RAFTER SPACING (inches)	SPECIES AND GRADE		DEAD LOAD = 10 pounds per square foot					DEAD LOAD = 20 pounds per square foot				
			2 × 4	2 × 6	2 × 8	2 × 10	2 × 12	2 × 4	2 × 6	2 × 8	2 × 10	2 × 12
			Maximum rafter spans									
			(ft. - in.)	(ft. - in.)	(ft. - in.)	(ft. - in.)	(ft. - in.)	(ft. - in.)	(ft. - in.)	(ft. - in.)	(ft. - in.)	(ft. - in.)
19.2	Douglas Fir-Larch	SS	9-10	15-5	20-4	25-11	Note a	9-10	14-7	18-6	22-7	Note a
	Douglas Fir-Larch	#1	9-5	14-0	17-9	21-8	25-2	8-4	12-2	15-4	18-9	21-9
	Douglas Fir-Larch	#2	8-11	13-1	16-7	20-3	23-6	7-9	11-4	14-4	17-7	20-4
	Douglas Fir-Larch	#3	6-9	9-11	12-7	15-4	17-9	5-10	8-7	10-10	13-3	15-5
	Hem-Fir	SS	9-3	14-7	19-2	24-6	Note a	9-3	14-4	18-2	22-3	25-9
	Hem-Fir	#1	9-1	13-8	17-4	21-1	24-6	8-1	11-10	15-0	18-4	21-3
	Hem-Fir	#2	8-8	12-11	16-4	20-0	23-2	7-8	11-2	14-2	17-4	20-1
	Hem-Fir	#3	6-9	9-11	12-7	15-4	17-9	5-10	8-7	10-10	13-3	15-5
	Southern Pine	SS	9-8	15-2	19-11	25-5	Note a	9-8	15-2	19-11	25-5	Note a
	Southern Pine	#1	9-5	14-10	19-7	23-7	Note a	9-3	13-8	17-2	20-5	24-4
	Southern Pine	#2	9-3	13-9	17-9	21-2	24-10	8-4	11-11	15-4	18-4	21-6
	Southern Pine	#3	7-3	10-8	13-7	16-0	19-1	6-3	9-3	11-9	13-10	16-6
	Spruce-Pine-Fir	SS	9-1	14-3	18-9	23-11	Note a	9-1	13-7	17-2	21-0	24-4
	Spruce-Pine-Fir	#1	8-10	13-1	16-7	20-3	23-6	7-9	11-4	14-4	17-7	20-4
	Spruce-Pine-Fir	#2	8-10	13-1	16-7	20-3	23-6	7-9	11-4	14-4	17-7	20-4
	Spruce-Pine-Fir	#3	6-9	9-11	12-7	15-4	17-9	5-10	8-7	10-10	13-3	15-5
24	Douglas Fir-Larch	SS	9-1	14-4	18-10	23-4	Note a	8-11	13-1	16-7	20-3	23-5
	Douglas Fir-Larch	#1	8-7	12-6	15-10	19-5	22-6	7-5	10-10	13-9	16-9	19-6
	Douglas Fir-Larch	#2	8-0	11-9	14-10	18-2	21-0	6-11	10-2	12-10	15-8	18-3
	Douglas Fir-Larch	#3	6-1	8-10	11-3	13-8	15-11	5-3	7-8	9-9	11-10	13-9
	Hem-Fir	SS	8-7	13-6	17-10	22-9	Note a	8-7	12-10	16-3	19-10	23-0
	Hem-Fir	#1	8-4	12-3	15-6	18-11	21-11	7-3	10-7	13-5	16-4	19-0
	Hem-Fir	#2	7-11	11-7	14-8	17-10	20-9	6-10	10-0	12-8	15-6	17-11
	Hem-Fir	#3	6-1	8-10	11-3	13-8	15-11	5-3	7-8	9-9	11-10	13-9
	Southern Pine	SS	8-11	14-1	18-6	23-8	Note a	8-11	14-1	18-6	22-11	Note a
	Southern Pine	#1	8-9	13-9	17-9	21-1	25-2	8-3	12-3	15-4	18-3	21-9
	Southern Pine	#2	8-7	12-3	15-10	18-11	22-2	7-5	10-8	13-9	16-5	19-3
	Southern Pine	#3	6-5	9-6	12-1	14-4	17-1	5-7	8-3	10-6	12-5	14-9
	Spruce-Pine-Fir	SS	8-5	13-3	17-5	21-8	25-2	8-4	12-2	15-4	18-9	21-9
	Spruce-Pine-Fir	#1	8-0	11-9	14-10	18-2	21-0	6-11	10-2	12-10	15-8	18-3
	Spruce-Pine-Fir	#2	8-0	11-9	14-10	18-2	21-0	6-11	10-2	12-10	15-8	18-3
	Spruce-Pine-Fir	#3	6-1	8-10	11-3	13-8	15-11	5-3	7-8	9-9	11-10	13-9

For SI: 1 inch = 25.4 mm, 1 foot = 304.8 mm, 1 pound per square foot = 47.9 N/m².

a. Span exceeds 26 feet in length. Check sources for availability of lumber in lengths greater than 20 feet.

TABLE 2308.10.3(2)
RAFTER SPANS FOR COMMON LUMBER SPECIES
(Roof Live Load = 20 pounds per square foot, Ceiling Not Attached to Rafters, $L/\Delta = 240$)

RAFTER SPACING (inches)	SPECIES AND GRADE		DEAD LOAD = 10 pounds per square foot					DEAD LOAD = 20 pounds per square foot				
			2 × 4	2 × 6	2 × 8	2 × 10	2 × 12	2 × 4	2 × 6	2 × 8	2 × 10	2 × 12
			Maximum rafter spans									
			(ft. - in.)	(ft. - in.)	(ft. - in.)	(ft. - in.)	(ft. - in.)	(ft. - in.)	(ft. - in.)	(ft. - in.)	(ft. - in.)	(ft. - in.)
12	Douglas Fir-Larch	SS	10-5	16-4	21-7	Note a	Note a	10-5	16-4	21-7	Note a	Note a
	Douglas Fir-Larch	#1	10-0	15-9	20-10	Note a	Note a	10-0	15-4	19-5	23-9	Note a
	Douglas Fir-Larch	#2	9-10	15-6	20-5	25-8	Note a	9-10	14-4	18-2	22-3	25-9
	Douglas Fir-Larch	#3	8-7	12-6	15-10	19-5	22-6	7-5	10-10	13-9	16-9	19-6
	Hem-Fir	SS	9-10	15-6	20-5	Note a	Note a	9-10	15-6	20-5	Note a	Note a
	Hem-Fir	#1	9-8	15-2	19-11	25-5	Note a	9-8	14-11	18-11	23-2	Note a
	Hem-Fir	#2	9-2	14-5	19-0	24-3	Note a	9-2	14-2	17-11	21-11	25-5
	Hem-Fir	#3	8-7	12-6	15-10	19-5	22-6	7-5	10-10	13-9	16-9	19-6
	Southern Pine	SS	10-3	16-1	21-2	Note a	Note a	10-3	16-1	21-2	Note a	Note a
	Southern Pine	#1	10-0	15-9	20-10	Note a	Note a	10-0	15-9	20-10	25-10	Note a
	Southern Pine	#2	9-10	15-6	20-5	Note a	Note a	9-10	15-1	19-5	23-2	Note a
	Southern Pine	#3	9-1	13-6	17-2	20-3	24-1	7-11	11-8	14-10	17-6	20-11
	Spruce-Pine-Fir	SS	9-8	15-2	19-11	25-5	Note a	9-8	15-2	19-11	25-5	Note a
	Spruce-Pine-Fir	#1	9-5	14-9	19-6	24-10	Note a	9-5	14-4	18-2	22-3	25-9
	Spruce-Pine-Fir	#2	9-5	14-9	19-6	24-10	Note a	9-5	14-4	18-2	22-3	25-9
	Spruce-Pine-Fir	#3	8-7	12-6	15-10	19-5	22-6	7-5	10-10	13-9	16-9	19-6
16	Douglas Fir-Larch	SS	9-6	14-11	19-7	25-0	Note a	9-6	14-11	19-7	24-9	Note a
	Douglas Fir-Larch	#1	9-1	14-4	18-11	23-9	Note a	9-1	13-3	16-10	20-7	23-10
	Douglas Fir-Larch	#2	8-11	14-1	18-2	22-3	25-9	8-6	12-5	15-9	19-3	22-4
	Douglas Fir-Larch	#3	7-5	10-10	13-9	16-9	19-6	6-5	9-5	11-11	14-6	16-10
	Hem-Fir	SS	8-11	14-1	18-6	23-8	Note a	8-11	14-1	18-6	23-8	Note a
	Hem-Fir	#1	8-9	13-9	18-1	23-1	Note a	8-9	12-11	16-5	20-0	23-3
	Hem-Fir	#2	8-4	13-1	17-3	21-11	25-5	8-4	12-3	15-6	18-11	22-0
	Hem-Fir	#3	7-5	10-10	13-9	16-9	19-6	6-5	9-5	11-11	14-6	16-10
	Southern Pine	SS	9-4	14-7	19-3	24-7	Note a	9-4	14-7	19-3	24-7	Note a
	Southern Pine	#1	9-1	14-4	18-11	24-1	Note a	9-1	14-4	18-10	22-4	Note a
	Southern Pine	#2	8-11	14-1	18-6	23-2	Note a	8-11	13-0	16-10	20-1	23-7
	Southern Pine	#3	7-11	11-8	14-10	17-6	20-11	6-10	10-1	12-10	15-2	18-1
	Spruce-Pine-Fir	SS	8-9	13-9	18-1	23-1	Note a	8-9	13-9	18-1	23-0	Note a
	Spruce-Pine-Fir	#1	8-7	13-5	17-9	22-3	25-9	8-6	12-5	15-9	19-3	22-4
	Spruce-Pine-Fir	#2	8-7	13-5	17-9	22-3	25-9	8-6	12-5	15-9	19-3	22-4
	Spruce-Pine-Fir	#3	7-5	10-10	13-9	16-9	19-6	6-5	9-5	11-11	14-6	16-10

(continued)

TABLE 2308.10.3(2)—continued
RAFTER SPANS FOR COMMON LUMBER SPECIES
(Roof Live Load = 20 pounds per square foot, Ceiling Not Attached to Rafters, $L/\Delta = 240$)

RAFTER SPACING (inches)	SPECIES AND GRADE		DEAD LOAD = 10 pounds per square foot					DEAD LOAD = 20 pounds per square foot				
			2 × 4	2 × 6	2 × 8	2 × 10	2 × 12	2 × 4	2 × 6	2 × 8	2 × 10	2 × 12
			\(ft.-in.\)	\(ft.-in.\)	\(ft.-in.\)	\(ft.-in.\)	\(ft.-in.\)	\(ft.-in.\)	\(ft.-in.\)	\(ft.-in.\)	\(ft.-in.\)	\(ft.-in.\)
19.2	Douglas Fir-Larch	SS	8-11	14-0	18-5	23-7	Note a	8-11	14-0	18-5	22-7	Note a
	Douglas Fir-Larch	#1	8-7	13-6	17-9	21-8	25-2	8-4	12-2	15-4	18-9	21-9
	Douglas Fir-Larch	#2	8-5	13-1	16-7	20-3	23-6	7-9	11-4	14-4	17-7	20-4
	Douglas Fir-Larch	#3	6-9	9-11	12-7	15-4	17-9	5-10	8-7	10-10	13-3	15-5
	Hem-Fir	SS	8-5	13-3	17-5	22-3	Note a	8-5	13-3	17-5	22-3	25-9
	Hem-Fir	#1	8-3	12-11	17-1	21-1	24-6	8-1	11-10	15-0	18-4	21-3
	Hem-Fir	#2	7-10	12-4	16-3	20-0	23-2	7-8	11-2	14-2	17-4	20-1
	Hem-Fir	#3	6-9	9-11	12-7	15-4	17-9	5-10	8-7	10-10	13-3	15-5
	Southern Pine	SS	8-9	13-9	18-1	23-1	Note a	8-9	13-9	18-1	23-1	Note a
	Southern Pine	#1	8-7	13-6	17-9	22-8	Note a	8-7	13-6	17-2	20-5	24-4
	Southern Pine	#2	8-5	13-3	17-5	21-2	24-10	8-4	11-11	15-4	18-4	21-6
	Southern Pine	#3	7-3	10-8	13-7	16-0	19-1	6-3	9-3	11-9	13-10	16-6
	Spruce-Pine-Fir	SS	8-3	12-11	17-1	21-9	Note a	8-3	12-11	17-1	21-0	24-4
	Spruce-Pine-Fir	#1	8-1	12-8	16-7	20-3	23-6	7-9	11-4	14-4	17-7	20-4
	Spruce-Pine-Fir	#2	8-1	12-8	16-7	20-3	23-6	7-9	11-4	14-4	17-7	20-4
	Spruce-Pine-Fir	#3	6-9	9-11	12-7	15-4	17-9	5-10	8-7	10-10	13-3	15-5
24	Douglas Fir-Larch	SS	8-3	13-0	17-2	21-10	Note a	8-3	13-0	16-7	20-3	23-5
	Douglas Fir-Larch	#1	8-0	12-6	15-10	19-5	22-6	7-5	10-10	13-9	16-9	19-6
	Douglas Fir-Larch	#2	7-10	11-9	14-10	18-2	21-0	6-11	10-2	12-10	15-8	18-3
	Douglas Fir-Larch	#3	6-1	8-10	11-3	13-8	15-11	5-3	7-8	9-9	11-10	13-9
	Hem-Fir	SS	7-10	12-3	16-2	20-8	25-1	7-10	12-3	16-2	19-10	23-0
	Hem-Fir	#1	7-8	12-0	15-6	18-11	21-11	7-3	10-7	13-5	16-4	19-0
	Hem-Fir	#2	7-3	11-5	14-8	17-10	20-9	6-10	10-0	12-8	15-6	17-11
	Hem-Fir	#3	6-1	8-10	11-3	13-8	15-11	5-3	7-8	9-9	11-10	13-9
	Southern Pine	SS	8-1	12-9	16-10	21-6	Note a	8-1	12-9	16-10	21-6	Note a
	Southern Pine	#1	8-0	12-6	16-6	21-1	25-2	8-0	12-3	15-4	18-3	21-9
	Southern Pine	#2	7-10	12-3	15-10	18-11	22-2	7-5	10-8	13-9	16-5	19-3
	Southern Pine	#3	6-5	9-6	12-1	14-4	17-1	5-7	8-3	10-6	12-5	14-9
	Spruce-Pine-Fir	SS	7-8	12-0	15-10	20-2	24-7	7-8	12-0	15-4	18-9	21-9
	Spruce-Pine-Fir	#1	7-6	11-9	14-10	18-2	21-0	6-11	10-2	12-10	15-8	18-3
	Spruce-Pine-Fir	#2	7-6	11-9	14-10	18-2	21-0	6-11	10-2	12-10	15-8	18-3
	Spruce-Pine-Fir	#3	6-1	8-10	11-3	13-8	15-11	5-3	7-8	9-9	11-10	13-9

For SI: 1 inch = 25.4 mm, 1 foot = 304.8 mm, 1 pound per square foot = 47.9 N/m².
a. Span exceeds 26 feet in length. Check sources for availability of lumber in lengths greater than 20 feet.

TABLE 2308.10.3(3)
RAFTER SPANS FOR COMMON LUMBER SPECIES
(Ground Snow Load = 30 pounds per square foot, Ceiling Not Attached to Rafters, L/Δ = 180)

RAFTER SPACING (inches)	SPECIES AND GRADE		DEAD LOAD = 10 pounds per square foot					DEAD LOAD = 20 pounds per square foot			
			2 × 4	2 × 6	2 × 8	2 × 10	2 × 12	2 × 6	2 × 8	2 × 10	2 × 12
			Maximum rafter spans								
			(ft. - in.)	(ft. - in.)	(ft. - in.)	(ft. - in.)	(ft. - in.)	(ft. - in.)	(ft. - in.)	(ft. - in.)	(ft. - in.)
12	Douglas Fir-Larch	SS	10-0	15-9	20-9	Note a	Note a	15-9	20-1	24-6	Note a
	Douglas Fir-Larch	#1	9-8	14-9	18-8	22-9	Note a	13-2	16-8	20-4	23-7
	Douglas Fir-Larch	#2	9-5	13-9	17-5	21-4	24-8	12-4	15-7	19-1	22-1
	Douglas Fir-Larch	#3	7-1	10-5	13-2	16-1	18-8	9-4	11-9	14-5	16-8
	Hem-Fir	SS	9-6	14-10	19-7	25-0	Note a	14-10	19-7	24-1	Note a
	Hem-Fir	#1	9-3	14-4	18-2	22-2	25-9	12-10	16-3	19-10	23-0
	Hem-Fir	#2	8-10	13-7	17-2	21-0	24-4	12-2	15-4	18-9	21-9
	Hem-Fir	#3	7-1	10-5	13-2	16-1	18-8	9-4	11-9	14-5	16-8
	Southern Pine	SS	9-10	15-6	20-5	Note a	Note a	15-6	20-5	Note a	Note a
	Southern Pine	#1	9-8	15-2	20-0	24-9	Note a	14-10	18-8	22-2	Note a
	Southern Pine	#2	9-6	14-5	18-8	22-3	Note a	12-11	16-8	19-11	23-4
	Southern Pine	#3	7-7	11-2	14-3	16-10	20-0	10-0	12-9	15-1	17-11
	Spruce-Pine-Fir	SS	9-3	14-7	19-2	24-6	Note a	14-7	18-8	22-9	Note a
	Spruce-Pine-Fir	#1	9-1	13-9	17-5	21-4	24-8	12-4	15-7	19-1	22-1
	Spruce-Pine-Fir	#2	9-1	13-9	17-5	21-4	24-8	12-4	15-7	19-1	22-1
	Spruce-Pine-Fir	#3	7-1	10-5	13-2	16-1	18-8	9-4	11-9	14-5	16-8
16	Douglas Fir-Larch	SS	9-1	14-4	18-10	23-9	Note a	13-9	17-5	21-3	24-8
	Douglas Fir-Larch	#1	8-9	12-9	16-2	19-9	22-10	11-5	14-5	17-8	20-5
	Douglas Fir-Larch	#2	8-2	11-11	15-1	18-5	21-5	10-8	13-6	16-6	19-2
	Douglas Fir-Larch	#3	6-2	9-0	11-5	13-11	16-2	8-1	10-3	12-6	14-6
	Hem-Fir	SS	8-7	13-6	17-10	22-9	Note a	13-6	17-1	20-10	24-2
	Hem-Fir	#1	8-5	12-5	15-9	19-3	22-3	11-1	14-1	17-2	19-11
	Hem-Fir	#2	8-0	11-9	14-11	18-2	21-1	10-6	13-4	16-3	18-10
	Hem-Fir	#3	6-2	9-0	11-5	13-11	16-2	8-1	10-3	12-6	14-6
	Southern Pine	SS	8-11	14-1	18-6	23-8	Note a	14-1	18-6	23-8	Note a
	Southern Pine	#1	8-9	13-9	18-1	21-5	25-7	12-10	16-2	19-2	22-10
	Southern Pine	#2	8-7	12-6	16-2	19-3	22-7	11-2	14-5	17-3	20-2
	Southern Pine	#3	6-7	9-8	12-4	14-7	17-4	8-8	11-0	13-0	15-6
	Spruce-Pine-Fir	SS	8-5	13-3	17-5	22-1	25-7	12-9	16-2	19-9	22-10
	Spruce-Pine-Fir	#1	8-2	11-11	15-1	18-5	21-5	10-8	13-6	16-6	19-2
	Spruce-Pine-Fir	#2	8-2	11-11	15-1	18-5	21-5	10-8	13-6	16-6	19-2
	Spruce-Pine-Fir	#3	6-2	9-0	11-5	13-11	16-2	8-1	10-3	12-6	14-6

(continued)

TABLE 2308.10.3(3)—continued
RAFTER SPANS FOR COMMON LUMBER SPECIES
(Ground Snow Load = 30 pounds per square foot, Ceiling Not Attached to Rafters, $L/\Delta = 180$)

Maximum rafter spans. All values are given in ft. - in.

RAFTER SPACING (inches)	SPECIES AND GRADE	Grade	DEAD LOAD = 10 pounds per square foot 2×4	2×6	2×8	2×10	2×12	DEAD LOAD = 20 pounds per square foot 2×4	2×6	2×8	2×10	2×12
19.2	Douglas Fir-Larch	SS	8-7	13-6	17-9	21-8	25-2	8-7	12-6	15-10	19-5	22-6
	Douglas Fir-Larch	#1	7-11	11-8	14-9	18-0	20-11	7-1	10-5	13-2	16-1	18-8
	Douglas Fir-Larch	#2	7-5	10-11	13-9	16-10	19-6	6-8	9-9	12-4	15-1	17-6
	Douglas Fir-Larch	#3	5-7	8-3	10-5	12-9	14-9	5-0	7-4	9-4	11-5	13-2
	Hem-Fir	SS	8-1	12-9	16-9	21-4	24-8	8-1	12-4	15-7	19-1	22-1
	Hem-Fir	#1	7-9	11-4	14-4	17-7	20-4	6-11	10-2	12-10	15-8	18-2
	Hem-Fir	#2	7-4	10-9	13-7	16-7	19-3	6-7	9-7	12-2	14-10	17-3
	Hem-Fir	#3	5-7	8-3	10-5	12-9	14-9	5-0	7-4	9-4	11-5	13-2
	Southern Pine	SS	8-5	13-3	17-5	22-3	Note a	8-5	13-3	17-5	22-0	25-9
	Southern Pine	#1	8-3	13-0	16-6	19-7	23-4	7-11	11-9	14-9	17-6	20-11
	Southern Pine	#2	7-11	11-5	14-9	17-7	20-7	7-1	10-2	13-2	15-9	18-5
	Southern Pine	#3	6-0	8-10	11-3	13-4	15-10	5-4	7-11	10-1	11-11	14-2
	Spruce-Pine-Fir	SS	7-11	12-5	16-5	20-2	23-4	7-11	11-8	14-9	18-0	20-11
	Spruce-Pine-Fir	#1	7-5	10-11	13-9	16-10	19-6	6-8	9-9	12-4	15-1	17-6
	Spruce-Pine-Fir	#2	7-5	10-11	13-9	16-10	19-6	6-8	9-9	12-4	15-1	17-6
	Spruce-Pine-Fir	#3	5-7	8-3	10-5	12-9	14-9	5-0	7-4	9-4	11-5	13-2
24	Douglas Fir-Larch	SS	7-11	12-6	15-10	19-5	22-6	7-8	11-3	14-2	17-4	20-1
	Douglas Fir-Larch	#1	7-1	10-5	13-2	16-1	18-8	6-4	9-4	11-9	14-5	16-8
	Douglas Fir-Larch	#2	6-8	9-9	12-4	15-1	17-6	5-11	8-8	11-0	13-6	15-7
	Douglas Fir-Larch	#3	5-0	7-4	9-4	11-5	13-2	4-6	6-7	8-4	10-2	11-10
	Hem-Fir	SS	7-6	11-10	15-7	19-1	22-1	7-6	11-0	13-11	17-0	19-9
	Hem-Fir	#1	6-11	10-2	12-10	15-8	18-2	6-2	9-1	11-6	14-0	16-3
	Hem-Fir	#2	6-7	9-7	12-2	14-10	17-3	5-10	8-7	10-10	13-3	15-5
	Hem-Fir	#3	5-0	7-4	9-4	11-5	13-2	4-6	6-7	8-4	10-2	11-10
	Southern Pine	SS	7-10	12-3	16-2	20-8	25-1	7-10	12-3	16-2	19-8	23-0
	Southern Pine	#1	7-8	11-9	14-9	17-6	20-11	7-1	10-6	13-2	15-8	18-8
	Southern Pine	#2	7-1	10-2	13-2	15-9	18-5	6-4	9-2	11-9	14-1	16-6
	Southern Pine	#3	5-4	7-11	10-1	11-11	14-2	4-9	7-1	9-0	10-8	12-8
	Spruce-Pine-Fir	SS	7-4	11-7	14-9	18-0	20-11	7-1	10-5	13-2	16-1	18-8
	Spruce-Pine-Fir	#1	6-8	9-9	12-4	15-1	17-6	5-11	8-8	11-0	13-6	15-7
	Spruce-Pine-Fir	#2	6-8	9-9	12-4	15-1	17-6	5-11	8-8	11-0	13-6	15-7
	Spruce-Pine-Fir	#3	5-0	7-4	9-4	11-5	13-2	4-6	6-7	8-4	10-2	11-10

For SI: 1 inch = 25.4 mm, 1 foot = 304.8 mm, 1 pound per square foot = 47.9 N/m².
a. Span exceeds 26 feet in length. Check sources for availability of lumber in lengths greater than 20 feet.

TABLE 2308.10.3(4)
RAFTER SPANS FOR COMMON LUMBER SPECIES
(Ground Snow Load = 50 pounds per square foot, Ceiling Not Attached to Rafters, $L/\Delta = 180$)

RAFTER SPACING (inches)	SPECIES AND GRADE		DEAD LOAD = 10 pounds per square foot					DEAD LOAD = 20 pounds per square foot				
			Maximum rafter spans									
			2 × 4	2 × 6	2 × 8	2 × 10	2 × 12	2 × 4	2 × 6	2 × 8	2 × 10	2 × 12
			(ft. - in.)	(ft. - in.)	(ft. - in.)	(ft. - in.)	(ft. - in.)	(ft. - in.)	(ft. - in.)	(ft. - in.)	(ft. - in.)	(ft. - in.)
12	Douglas Fir-Larch	SS	8-5	13-3	17-6	22-4	26-0	8-5	13-3	17-0	20-9	24-10
	Douglas Fir-Larch	#1	8-2	12-0	15-3	18-7	21-7	7-7	11-2	14-1	17-3	20-0
	Douglas Fir-Larch	#2	7-8	11-3	14-3	17-5	20-2	7-1	10-5	13-2	16-1	18-8
	Douglas Fir-Larch	#3	5-10	8-6	10-9	13-2	15-3	5-5	7-10	10-0	12-2	14-1
	Hem-Fir	SS	8-0	12-6	16-6	21-1	25-6	8-0	12-6	16-6	20-4	23-7
	Hem-Fir	#1	7-10	11-9	14-10	18-1	21-0	7-5	10-10	13-9	16-9	19-5
	Hem-Fir	#2	7-5	11-1	14-0	17-2	19-11	7-0	10-3	13-0	15-10	18-5
	Hem-Fir	#3	5-10	8-6	10-9	13-2	15-3	5-5	7-10	10-0	12-2	14-1
	Southern Pine	SS	8-4	13-0	17-2	21-11	Note a	8-4	13-0	17-2	21-11	Note a
	Southern Pine	#1	8-2	12-10	16-10	20-3	24-1	8-2	12-6	15-9	18-9	22-4
	Southern Pine	#2	8-0	11-9	15-3	18-2	21-3	7-7	10-11	14-1	16-10	19-9
	Southern Pine	#3	6-2	9-2	11-8	13-9	16-4	5-9	8-5	10-9	12-9	15-2
	Spruce-Pine-Fir	SS	7-10	12-3	16-2	20-8	24-1	7-10	12-3	15-9	19-3	22-4
	Spruce-Pine-Fir	#1	7-8	11-3	14-3	17-5	20-2	7-1	10-5	13-2	16-1	18-8
	Spruce-Pine-Fir	#2	7-8	11-3	14-3	17-5	20-2	7-1	10-5	13-2	16-1	18-8
	Spruce-Pine-Fir	#3	5-10	8-6	10-9	13-2	15-3	5-5	7-10	10-0	12-2	14-1
16	Douglas Fir-Larch	SS	7-8	12-1	15-10	19-5	22-6	7-8	11-7	14-8	17-11	20-10
	Douglas Fir-Larch	#1	7-1	10-5	13-2	16-1	18-8	6-7	9-8	12-2	14-11	17-3
	Douglas Fir-Larch	#2	6-8	9-9	12-4	15-1	17-6	6-2	9-0	11-5	13-11	16-2
	Douglas Fir-Larch	#3	5-0	7-4	9-4	11-5	13-2	4-8	6-10	8-8	10-6	12-3
	Hem-Fir	SS	7-3	11-5	15-0	19-1	22-1	7-3	11-5	14-5	17-8	20-5
	Hem-Fir	#1	6-11	10-2	12-10	15-8	18-2	6-5	9-5	11-11	14-6	16-10
	Hem-Fir	#2	6-7	9-7	12-2	14-10	17-3	6-1	8-11	11-3	13-9	15-11
	Hem-Fir	#3	5-0	7-4	9-4	11-5	13-2	4-8	6-10	8-8	10-6	12-3
	Southern Pine	SS	7-6	11-10	15-7	19-11	24-3	7-6	11-10	15-7	19-11	23-10
	Southern Pine	#1	7-5	11-7	14-9	17-6	20-11	7-4	10-10	13-8	16-2	19-4
	Southern Pine	#2	7-1	10-2	13-2	15-9	18-5	6-7	9-5	12-2	14-7	17-1
	Southern Pine	#3	5-4	7-11	10-1	11-11	14-2	4-11	7-4	9-4	11-0	13-1
	Spruce-Pine-Fir	SS	7-1	11-2	14-8	18-0	20-11	7-1	10-9	13-8	16-8	19-4
	Spruce-Pine-Fir	#1	6-8	9-9	12-4	15-1	17-6	6-2	9-0	11-5	13-11	16-2
	Spruce-Pine-Fir	#2	6-8	9-9	12-4	15-1	17-6	6-2	9-0	11-5	13-11	16-2
	Spruce-Pine-Fir	#3	5-0	7-4	9-4	11-5	13-2	4-8	6-10	8-8	10-6	12-3

(continued)

TABLE 2308.10.3(4)—continued
RAFTER SPANS FOR COMMON LUMBER SPECIES
(Ground Snow Load = 50 pounds per square foot, Ceiling Not Attached to Rafters, $L/\Delta = 180$)

RAFTER SPACING (inches)	SPECIES AND GRADE		DEAD LOAD = 10 pounds per square foot					DEAD LOAD = 20 pounds per square foot				
			2 × 4	2 × 6	2 × 8	2 × 10	2 × 12	2 × 4	2 × 6	2 × 8	2 × 10	2 × 12
			Maximum rafter spans									
			(ft. - in.)	(ft. - in.)	(ft. - in.)	(ft. - in.)	(ft. - in.)	(ft. - in.)	(ft. - in.)	(ft. - in.)	(ft. - in.)	(ft. - in.)
19.2	Douglas Fir-Larch	SS	7-3	11-4	14-6	17-8	20-6	7-3	10-7	13-5	16-5	19-0
	Douglas Fir-Larch	#1	6-6	9-6	12-0	14-8	17-1	6-0	8-10	11-2	13-7	15-9
	Douglas Fir-Larch	#2	6-1	8-11	11-3	13-9	15-11	5-7	8-3	10-5	12-9	14-9
	Douglas Fir-Larch	#3	4-7	6-9	8-6	10-5	12-1	4-3	6-3	7-11	9-7	11-2
	Hem-Fir	SS	6-10	10-9	14-2	17-5	20-2	6-10	10-5	13-2	16-1	18-8
	Hem-Fir	#1	6-4	9-3	11-9	14-4	16-7	5-10	8-7	10-10	13-3	15-5
	Hem-Fir	#2	6-0	8-9	11-1	13-7	15-9	5-7	8-1	10-3	12-7	14-7
	Hem-Fir	#3	4-7	6-9	8-6	10-5	12-1	4-3	6-3	7-11	9-7	11-2
	Southern Pine	SS	7-1	11-2	14-8	18-9	22-10	7-1	11-2	14-8	18-7	21-9
	Southern Pine	#1	7-0	10-8	13-5	16-0	19-1	6-8	9-11	12-5	14-10	17-8
	Southern Pine	#2	6-6	9-4	12-0	14-4	16-10	6-0	8-8	11-2	13-4	15-7
	Southern Pine	#3	4-11	7-3	9-2	10-10	12-11	4-6	6-8	8-6	10-1	12-0
	Spruce-Pine-Fir	SS	6-8	10-6	13-5	16-5	19-1	6-8	9-10	12-5	15-3	17-8
	Spruce-Pine-Fir	#1	6-1	8-11	11-3	13-9	15-11	5-7	8-3	10-5	12-9	14-9
	Spruce-Pine-Fir	#2	6-1	8-11	11-3	13-9	15-11	5-7	8-3	10-5	12-9	14-9
	Spruce-Pine-Fir	#3	4-7	6-9	8-6	10-5	12-1	4-3	6-3	7-11	9-7	11-2
24	Douglas Fir-Larch	SS	6-8	10-3	13-0	15-10	18-4	6-6	9-6	12-0	14-8	17-0
	Douglas Fir-Larch	#1	5-10	8-6	10-9	13-2	15-3	5-5	7-10	10-0	12-2	14-1
	Douglas Fir-Larch	#2	5-5	7-11	10-1	12-4	14-3	5-0	7-4	9-4	11-5	13-2
	Douglas Fir-Larch	#3	4-1	6-0	7-7	9-4	10-9	3-10	5-7	7-1	8-7	10-0
	Hem-Fir	SS	6-4	9-11	12-9	15-7	18-0	6-4	9-4	11-9	14-5	16-8
	Hem-Fir	#1	5-8	8-3	10-6	12-10	14-10	5-3	7-8	9-9	11-10	13-9
	Hem-Fir	#2	5-4	7-10	9-11	12-1	14-1	4-11	7-3	9-2	11-3	13-0
	Hem-Fir	#3	4-1	6-0	7-7	9-4	10-9	3-10	5-7	7-1	8-7	10-0
	Southern Pine	SS	6-7	10-4	13-8	17-5	21-0	6-7	10-4	13-8	16-7	19-5
	Southern Pine	#1	6-5	9-7	12-0	14-4	17-1	6-0	8-10	11-2	13-3	15-9
	Southern Pine	#2	5-10	8-4	10-9	12-10	15-1	5-5	7-9	10-0	11-11	13-11
	Southern Pine	#3	4-4	6-5	8-3	9-9	11-7	4-1	6-0	7-7	9-0	10-8
	Spruce-Pine-Fir	SS	6-2	9-6	12-0	14-8	17-1	6-0	8-10	11-2	13-7	15-9
	Spruce-Pine-Fir	#1	5-5	7-11	10-1	12-4	14-3	5-0	7-4	9-4	11-5	13-2
	Spruce-Pine-Fir	#2	5-5	7-11	10-1	12-4	14-3	5-0	7-4	9-4	11-5	13-2
	Spruce-Pine-Fir	#3	4-1	6-0	7-7	9-4	10-9	3-10	5-7	7-1	8-7	10-0

For SI: 1 inch = 25.4 mm, 1 foot = 304.8 mm, 1 pound per square foot = 47.9 N/m².

a. Span exceeds 26 feet in length. Check sources for availability of lumber in lengths greater than 20 feet.

TABLE 2308.10.3(5)
RAFTER SPANS FOR COMMON LUMBER SPECIES
(Ground Snow Load = 30 pounds per square foot, Ceiling Attached to Rafters, $L/\Delta = 240$)

RAFTER SPACING (inches)	SPECIES AND GRADE		DEAD LOAD = 10 pounds per square foot					DEAD LOAD = 20 pounds per square foot				
			2 × 4	2 × 6	2 × 8	2 × 10	2 × 12	2 × 4	2 × 6	2 × 8	2 × 10	2 × 12
			\multicolumn{10}{c}{Maximum rafter spans}									
			(ft. - in.)	(ft. - in.)	(ft. - in.)	(ft. - in.)	(ft. - in.)	(ft. - in.)	(ft. - in.)	(ft. - in.)	(ft. - in.)	(ft. - in.)
12	Douglas Fir-Larch	SS	9-1	14-4	18-10	24-1	Note a	9-1	14-4	18-10	24-1	Note a
	Douglas Fir-Larch	#1	8-9	13-9	18-2	22-9	Note a	8-9	13-2	16-8	20-4	23-7
	Douglas Fir-Larch	#2	8-7	13-6	17-5	21-4	24-8	8-5	12-4	15-7	19-1	22-1
	Douglas Fir-Larch	#3	7-1	10-5	13-2	16-1	18-8	6-4	9-4	11-9	14-5	16-8
	Hem-Fir	SS	8-7	13-6	17-10	22-9	Note a	8-7	13-6	17-10	22-9	Note a
	Hem-Fir	#1	8-5	13-3	17-5	22-2	25-9	8-5	12-10	16-3	19-10	23-0
	Hem-Fir	#2	8-0	12-7	16-7	21-0	24-4	8-0	12-2	15-4	18-9	21-9
	Hem-Fir	#3	7-1	10-5	13-2	16-1	18-8	6-4	9-4	11-9	14-5	16-8
	Southern Pine	SS	8-11	14-1	18-6	23-8	Note a	8-11	14-1	18-6	23-8	Note a
	Southern Pine	#1	8-9	13-9	18-2	23-2	Note a	8-9	13-9	18-2	22-2	Note a
	Southern Pine	#2	8-7	13-6	17-10	22-3	Note a	8-7	12-11	16-8	19-11	23-4
	Southern Pine	#3	7-7	11-2	14-3	16-10	20-0	6-9	10-0	12-9	15-1	17-11
	Spruce-Pine-Fir	SS	8-5	13-3	17-5	22-3	Note a	8-5	13-3	17-5	22-3	Note a
	Spruce-Pine-Fir	#1	8-3	12-11	17-0	21-4	24-8	8-3	12-4	15-7	19-1	22-1
	Spruce-Pine-Fir	#2	8-3	12-11	17-0	21-4	24-8	8-3	12-4	15-7	19-1	22-1
	Spruce-Pine-Fir	#3	7-1	10-5	13-2	16-1	18-8	6-4	9-4	11-9	14-5	16-8
16	Douglas Fir-Larch	SS	8-3	13-0	17-2	21-10	Note a	8-3	13-0	17-2	21-3	24-8
	Douglas Fir-Larch	#1	8-0	12-6	16-2	19-9	22-10	7-10	11-5	14-5	17-8	20-5
	Douglas Fir-Larch	#2	7-10	12-3	15-1	18-5	21-5	7-3	10-8	13-6	16-6	19-2
	Douglas Fir-Larch	#3	6-2	9-0	11-5	13-11	16-2	5-6	8-1	10-3	12-6	14-6
	Hem-Fir	SS	7-10	12-3	16-2	20-8	25-1	7-10	12-3	16-2	20-8	24-2
	Hem-Fir	#1	7-8	12-0	15-9	19-3	22-3	7-7	11-1	14-1	17-2	19-11
	Hem-Fir	#2	7-3	11-5	14-11	18-2	21-1	7-2	10-6	13-4	16-3	18-10
	Hem-Fir	#3	6-2	9-0	11-5	13-11	16-2	5-6	8-1	10-3	12-6	14-6
	Southern Pine	SS	8-1	12-9	16-10	21-6	Note a	8-1	12-9	16-10	21-6	Note a
	Southern Pine	#1	8-0	12-6	16-6	21-1	25-7	8-0	12-6	16-2	19-2	22-10
	Southern Pine	#2	7-10	12-3	16-2	19-3	22-7	7-10	11-2	14-5	17-3	20-2
	Southern Pine	#3	6-7	9-8	12-4	14-7	17-4	5-10	8-8	11-0	13-0	15-6
	Spruce-Pine-Fir	SS	7-8	12-0	15-10	20-2	24-7	7-8	12-0	15-10	19-9	22-10
	Spruce-Pine-Fir	#1	7-6	11-9	15-1	18-5	21-5	7-3	10-8	13-6	16-6	19-2
	Spruce-Pine-Fi	#2	7-6	11-9	15-1	18-5	21-5	7-3	10-8	13-6	16-6	19-2
	Spruce-Pine-Fi	#3	6-2	9-0	11-5	13-11	16-2	5-6	8-1	10-3	12-6	14-6

(continued)

TABLE 2308.10.3(5)—continued
RAFTER SPANS FOR COMMON LUMBER SPECIES
(Ground Snow Load = 30 pounds per square foot, Ceiling Attached to Rafters, L/Δ = 240)

RAFTER SPACING (inches)	SPECIES AND GRADE		DEAD LOAD = 10 pounds per square foot					DEAD LOAD = 20 pounds per square foot				
			Maximum rafter spans									
			2 × 4	2 × 6	2 × 8	2 × 10	2 × 12	2 × 4	2 × 6	2 × 8	2 × 10	2 × 12
			(ft.-in.)	(ft.-in.)	(ft.-in.)	(ft.-in.)	(ft.-in.)	(ft.-in.)	(ft.-in.)	(ft.-in.)	(ft.-in.)	(ft.-in.)
19.2	Douglas Fir-Larch	SS	7-9	12-3	16-1	20-7	25-0	7-9	12-3	15-10	19-5	22-6
	Douglas Fir-Larch	#1	7-6	11-8	14-9	18-0	20-11	7-1	10-5	13-2	16-1	18-8
	Douglas Fir-Larch	#2	7-4	10-11	13-9	16-10	19-6	6-8	9-9	12-4	15-1	17-6
	Douglas Fir-Larch	#3	5-7	8-3	10-5	12-9	14-9	5-0	7-4	9-4	11-5	13-2
	Hem-Fir	SS	7-4	11-7	15-3	19-5	23-7	7-4	11-7	15-3	19-1	22-1
	Hem-Fir	#1	7-2	11-4	14-4	17-7	20-4	6-11	10-2	12-10	15-8	18-2
	Hem-Fir	#2	6-10	10-9	13-7	16-7	19-3	6-7	9-7	12-2	14-10	17-3
	Hem-Fir	#3	5-7	8-3	10-5	12-9	14-9	5-0	7-4	9-4	11-5	13-2
	Southern Pine	SS	7-8	12-0	15-10	20-2	24-7	7-8	12-0	15-10	20-2	24-7
	Southern Pine	#1	7-6	11-9	15-6	19-7	23-4	7-6	11-9	14-9	17-6	20-11
	Southern Pine	#2	7-4	11-5	14-9	17-7	20-7	7-1	10-2	13-2	15-9	18-5
	Southern Pine	#3	6-0	8-10	11-3	13-4	15-10	5-4	7-11	10-1	11-11	14-2
	Spruce-Pine-Fir	SS	7-2	11-4	14-11	19-0	23-1	7-2	11-4	14-9	18-0	20-11
	Spruce-Pine-Fir	#1	7-0	10-11	13-9	16-10	19-6	6-8	9-9	12-4	15-1	17-6
	Spruce-Pine-Fir	#2	7-0	10-11	13-9	16-10	19-6	6-8	9-9	12-4	15-1	17-6
	Spruce-Pine-Fir	#3	5-7	8-3	10-5	12-9	14-9	5-0	7-4	9-4	11-5	13-2
24	Douglas Fir-Larch	SS	7-3	11-4	15-0	19-1	22-6	7-3	11-3	14-2	17-4	20-1
	Douglas Fir-Larch	#1	7-0	10-5	13-2	16-1	18-8	6-4	9-4	11-9	14-5	16-8
	Douglas Fir-Larch	#2	6-8	9-9	12-4	15-1	17-6	5-11	8-8	11-0	13-6	15-7
	Douglas Fir-Larch	#3	5-0	7-4	9-4	11-5	13-2	4-6	6-7	8-4	10-2	11-10
	Hem-Fir	SS	6-10	10-9	14-2	18-0	21-11	6-10	10-9	13-11	17-0	19-9
	Hem-Fir	#1	6-8	10-2	12-10	15-8	18-2	6-2	9-1	11-6	14-0	16-3
	Hem-Fir	#2	6-6	9-7	12-2	14-10	17-3	5-10	8-7	10-10	13-3	15-5
	Hem-Fir	#3	5-0	7-4	9-4	11-5	13-2	4-6	6-7	8-4	10-2	11-10
	Southern Pine	SS	7-1	11-2	14-8	18-9	22-10	7-1	11-2	14-8	18-9	22-10
	Southern Pine	#1	7-0	10-11	14-5	17-6	20-11	7-0	10-6	13-2	15-8	20-11
	Southern Pine	#2	6-10	10-2	13-2	15-9	18-5	6-4	9-2	11-9	14-1	16-6
	Southern Pine	#3	5-4	7-11	10-1	11-11	14-2	4-9	7-1	9-0	10-8	12-8
	Spruce-Pine-Fir	SS	6-8	10-6	13-10	17-8	20-11	6-8	10-5	13-2	16-1	18-8
	Spruce-Pine-Fir	#1	6-6	9-9	12-4	15-1	17-6	5-11	8-8	11-0	13-6	15-7
	Spruce-Pine-Fir	#2	6-6	9-9	12-4	15-1	17-6	5-11	8-8	11-0	13-6	15-7
	Spruce-Pine-Fir	#3	5-0	7-4	9-4	11-5	13-2	4-6	6-7	8-4	10-2	11-10

For SI: 1 inch = 25.4 mm, 1 foot = 304.8 mm, 1 pound per square foot = 47.9 N/m².

a. Span exceeds 26 feet in length. Check sources for availability of lumber in lengths greater than 20 feet.

TABLE 2308.10.3(6)
RAFTER SPANS FOR COMMON LUMBER SPECIES
(Ground Snow Load = 50 pounds per square foot, Ceiling Attached to Rafters, $L/\Delta = 240$)

RAFTER SPACING (inches)	SPECIES AND GRADE		DEAD LOAD = 10 pounds per square foot					DEAD LOAD = 20 pounds per square foot				
			2 × 4	2 × 6	2 × 8	2 × 10	2 × 12	2 × 4	2 × 6	2 × 8	2 × 10	2 × 12
			(ft.-in.)	(ft.-in.)	(ft.-in.)	(ft.-in.)	(ft.-in.)	(ft.-in.)	(ft.-in.)	(ft.-in.)	(ft.-in.)	(ft.-in.)
			Maximum rafter spans									
12	Douglas Fir-Larch	SS	7-8	12-1	15-11	20-3	24-8	7-8	12-1	15-11	20-3	24-0
	Douglas Fir-Larch	#1	7-5	11-7	15-3	18-7	21-7	7-5	11-2	14-1	17-3	20-0
	Douglas Fir-Larch	#2	7-3	11-3	14-3	17-5	20-2	7-1	10-5	13-2	16-1	18-8
	Douglas Fir-Larch	#3	5-10	8-6	10-9	13-2	15-3	5-5	7-10	10-0	12-2	14-1
	Hem-Fir	SS	7-3	11-5	15-0	19-2	23-4	7-3	11-5	15-0	19-2	23-4
	Hem-Fir	#1	7-1	11-2	14-8	18-1	21-0	7-1	10-10	13-9	16-9	19-5
	Hem-Fir	#2	6-9	10-8	14-0	17-2	19-11	6-9	10-3	13-0	15-10	18-5
	Hem-Fir	#3	5-10	8-6	10-9	13-2	15-3	5-5	7-10	10-0	12-2	14-1
	Southern Pine	SS	7-6	11-0	15-7	19-11	24-3	7-6	11-10	15-7	19-11	24-3
	Southern Pine	#1	7-5	11-7	15-4	19-7	23-9	7-5	11-7	15-4	18-9	22-4
	Southern Pine	#2	7-3	11-5	15-0	18-2	21-3	7-3	10-11	14-1	16-10	19-9
	Southern Pine	#3	6-2	9-2	11-8	13-9	16-4	5-9	8-5	10-9	12-9	15-2
	Spruce-Pine-Fir	SS	7-1	11-2	14-8	18-9	22-10	7-1	11-2	14-8	18-9	22-4
	Spruce-Pine-Fir	#1	6-11	10-11	14-3	17-5	20-2	6-11	10-5	13-2	16-1	18-8
	Spruce-Pine-Fir	#2	6-11	10-11	14-3	17-5	20-2	6-11	10-5	13-2	16-1	18-8
	Spruce-Pine-Fir	#3	5-10	8-6	10-9	13-2	15-3	5-5	7-10	10-0	12-2	14-1
16	Douglas Fir-Larch	SS	7-0	11-0	14-5	18-5	22-5	7-0	11-0	14-5	17-11	20-10
	Douglas Fir-Larch	#1	6-9	10-5	13-2	16-1	18-8	6-7	9-8	12-2	14-11	17-3
	Douglas Fir-Larch	#2	6-7	9-9	12-4	15-1	17-6	6-2	9-0	11-5	13-11	16-2
	Douglas Fir-Larch	#3	5-0	7-4	9-4	11-5	13-2	4-8	6-10	8-8	10-6	12-3
	Hem-Fir	SS	6-7	10-4	13-8	17-5	21-2	6-7	10-4	13-8	17-5	20-5
	Hem-Fir	#1	6-5	10-2	12-10	15-8	18-2	6-5	9-5	11-11	14-6	16-10
	Hem-Fir	#2	6-2	9-7	12-2	14-10	17-3	6-1	8-11	11-3	13-9	15-11
	Hem-Fir	#3	5-0	7-4	9-4	11-5	13-2	4-8	6-10	8-8	10-6	12-3
	Southern Pine	SS	6-10	10-9	14-2	18-1	22-0	6-10	10-9	14-2	18-1	22-0
	Southern Pine	#1	6-9	10-7	13-11	17-6	20-11	6-9	10-7	13-8	16-2	19-4
	Southern Pine	#2	6-7	10-2	13-2	15-9	18-5	6-7	9-5	12-2	14-7	17-1
	Southern Pine	#3	5-4	7-11	10-1	11-11	14-2	4-11	7-4	9-4	11-0	13-1
	Spruce-Pine-Fir	SS	6-5	10-2	13-4	17-0	20-9	6-5	10-2	13-4	16-8	19-4
	Spruce-Pine-Fir	#1	6-4	9-9	12-4	15-1	17-6	6-2	9-0	11-5	13-11	16-2
	Spruce-Pine-Fir	#2	6-4	9-9	12-4	15-1	17-6	6-2	9-0	11-5	13-11	16-2
	Spruce-Pine-Fir	#3	5-0	7-4	9-4	11-5	13-2	4-8	6-10	8-8	10-6	12-3

(continued)

47

TABLE 2308.10.3(6)—continued
RAFTER SPANS FOR COMMON LUMBER SPECIES
(Ground Snow Load = 50 pounds per square foot, Ceiling Attached to Rafters, $L/\Delta = 240$)

RAFTER SPACING (inches)	SPECIES AND GRADE		DEAD LOAD = 10 pounds per square foot					DEAD LOAD = 20 pounds per square foot				
			2 × 4	2 × 6	2 × 8	2 × 10	2 × 12	2 × 4	2 × 6	2 × 8	2 × 10	2 × 12
			Maximum rafter spans									
			(ft. - in.)	(ft. - in.)	(ft. - in.)	(ft. - in.)	(ft. - in.)	(ft. - in.)	(ft. - in.)	(ft. - in.)	(ft. - in.)	(ft. - in.)
19.2	Douglas Fir-Larch	SS	6-7	10-4	13-7	17-4	20-6	6-7	10-4	13-5	16-5	19-0
	Douglas Fir-Larch	#1	6-4	9-6	12-0	14-8	17-1	6-0	8-10	11-2	13-7	15-9
	Douglas Fir-Larch	#2	6-1	8-11	11-3	13-9	15-11	5-7	8-3	10-5	12-9	14-9
	Douglas Fir-Larch	#3	4-7	6-9	8-6	10-5	12-1	4-3	6-3	7-11	9-7	11-2
	Hem-Fir	SS	6-2	9-9	12-10	16-5	19-11	6-2	9-9	12-10	16-1	18-8
	Hem-Fir	#1	6-1	9-3	11-9	14-4	16-7	5-10	8-7	10-10	13-3	15-5
	Hem-Fir	#2	5-9	8-9	11-1	13-7	15-9	5-7	8-1	10-3	12-7	14-7
	Hem-Fir	#3	4-7	6-9	8-6	10-5	12-1	4-3	6-3	7-11	9-7	11-2
	Southern Pine	SS	6-5	10-2	13-4	17-0	20-9	6-5	10-2	13-4	17-0	20-9
	Southern Pine	#1	6-4	9-11	13-1	16-0	19-1	6-4	9-11	12-5	14-10	17-8
	Southern Pine	#2	6-2	9-4	12-0	14-4	16-10	6-0	8-8	11-2	13-4	15-7
	Southern Pine	#3	4-11	7-3	9-2	10-10	12-11	4-6	6-8	8-6	10-1	12-0
	Spruce-Pine-Fir	SS	6-1	9-6	12-7	16-0	19-1	6-1	9-6	12-5	15-3	17-8
	Spruce-Pine-Fir	#1	5-11	8-11	11-3	13-9	15-11	5-7	8-3	10-5	12-9	14-9
	Spruce-Pine-Fir	#2	5-11	8-11	11-3	13-9	15-11	5-7	8-3	10-5	12-9	14-9
	Spruce-Pine-Fir	#3	4-7	6-9	8-6	10-5	12-1	4-3	6-3	7-11	9-7	11-2
24	Douglas Fir-Larch	SS	6-1	9-7	12-7	15-10	18-4	6-1	9-6	12-0	14-8	17-0
	Douglas Fir-Larch	#1	5-10	8-6	10-9	13-2	15-3	5-5	7-10	10-0	12-2	14-1
	Douglas Fir-Larch	#2	5-5	7-11	10-1	12-4	14-3	5-0	7-4	9-4	11-5	13-2
	Douglas Fir-Larch	#3	4-1	6-0	7-7	9-4	10-9	3-10	5-7	7-1	8-7	10-0
	Hem-Fir	SS	5-9	9-1	11-11	15-2	18-0	5-9	9-1	11-9	14-5	16-8
	Hem-Fir	#1	5-8	8-3	10-6	12-10	14-10	5-3	7-8	9-9	11-10	13-9
	Hem-Fir	#2	5-4	7-10	9-11	12-1	14-1	4-11	7-3	9-2	11-3	13-0
	Hem-Fir	#3	4-1	6-0	7-7	9-4	10-9	3-10	5-7	7-1	8-7	10-0
	Southern Pine	SS	6-0	9-5	12-5	15-10	19-3	6-0	9-5	12-5	15-10	19-3
	Southern Pine	#1	5-10	9-3	12-0	14-4	17-1	5-10	8-10	11-2	13-3	15-9
	Southern Pine	#2	5-9	8-4	10-9	12-10	15-1	5-5	7-9	10-0	11-11	13-11
	Southern Pine	#3	4-4	6-5	8-3	9-9	11-7	4-1	6-0	7-7	9-0	10-8
	Spruce-Pine-Fir	SS	5-8	8-10	11-8	14-8	17-1	5-8	8-10	11-2	13-7	15-9
	Spruce-Pine-Fir	#1	5-5	7-11	10-1	12-4	14-3	5-0	7-4	9-4	11-5	13-2
	Spruce-Pine-Fir	#2	5-5	7-11	10-1	12-4	14-3	5-0	7-4	9-4	11-5	13-2
	Spruce-Pine-Fir	#3	4-1	6-0	7-7	9-4	10-9	3-10	5-7	7-1	8-7	10-0

For SI: 1 inch = 25.4 mm, 1 foot = 304.8 mm, 1 pound per square foot = 47.9 N/m².

TABLE 2308.10.4.1
RAFTER TIE CONNECTIONS[g]

RAFTER SLOPE	TIE SPACING (inches)	NO SNOW LOAD				GROUND SNOW LOAD (pound per square foot)							
						30 pounds per square foot				50 pounds per square foot			
						Roof span (feet)							
		12	20	28	36	12	20	28	36	12	20	28	36
		Required number of 16d common ($3^1/_2''$ x 0.162") nails[a,b] per connection[c,d,e,f]											
3:12	12	4	6	8	10	4	6	8	11	5	8	12	15
	16	5	7	10	13	5	8	11	14	6	11	15	20
	24	7	11	15	19	7	11	16	21	9	16	23	30
	32	10	14	19	25	10	16	22	28	12	27	30	40
	48	14	21	29	37	14	32	36	42	18	32	46	60
4:12	12	3	4	5	6	3	5	6	8	4	6	9	11
	16	3	5	7	8	4	6	8	11	5	8	12	15
	24	4	7	10	12	5	9	12	16	7	12	17	22
	32	6	9	13	16	8	12	16	22	10	16	24	30
	48	8	14	19	24	10	18	24	32	14	24	34	44
5:12	12	3	3	4	5	3	4	5	7	3	5	7	9
	16	3	4	5	7	3	5	7	9	4	7	9	12
	24	4	6	8	10	4	7	10	13	6	10	14	18
	32	5	8	10	13	6	10	14	18	8	14	18	24
	48	7	11	15	20	8	14	20	26	12	20	28	36
7:12	12	3	3	3	4	3	3	4	5	3	4	5	7
	16	3	3	4	5	3	4	5	6	3	5	7	9
	24	3	4	6	7	3	5	7	9	4	7	10	13
	32	4	6	8	10	4	8	10	12	6	10	14	18
	48	5	8	11	14	6	10	14	18	9	14	20	26
9:12	12	3	3	3	3	3	3	3	4	3	3	4	5
	16	3	3	3	4	3	3	4	5	3	4	5	7
	24	3	3	5	6	3	4	6	7	3	6	8	10
	32	3	4	6	8	4	6	8	10	5	8	10	14
	48	4	6	9	11	5	8	12	14	7	12	16	20
12:12	12	3	3	3	3	3	3	3	3	3	3	3	4
	16	3	3	3	3	3	3	3	4	3	3	4	5
	24	3	3	3	4	3	3	4	6	3	4	6	8
	32	3	3	4	5	3	5	6	8	4	6	8	10
	48	3	4	6	7	4	7	8	12	6	8	12	16

For SI: 1 inch = 25.4 mm, 1 foot = 304.8 mm, 1 pound per square foot = 47.8 N/m^2.

a. 40d box (5" × 0.162") or l6d sinker ($3^1/_4''$ × 0.148") nails are permitted to be substituted for 16d common ($3^1/_2''$ × 0.16") nails.

b. Nailing requirements are permitted to be reduced 25 percent if nails are clinched.

c. Rafter tie heel joint connections are not required where the ridge is supported by a load-bearing wall, header or ridge beam.

d. When intermediate support of the rafter is provided by vertical struts or purlins to a load-bearing wall, the tabulated heel joint connection requirements are permitted to be reduced proportionally to the reduction in span.

e. Equivalent nailing patterns are required for ceiling joist to ceiling joist lap splices.

f. Connected members shall be of sufficient size to prevent splitting due to nailing.

g. For snow loads less than 30 pounds per square foot, the required number of nails is permitted to be reduced by multiplying by the ratio of actual snow load plus 10 divided by 40, but not less than the number required for no snow load.

Figures

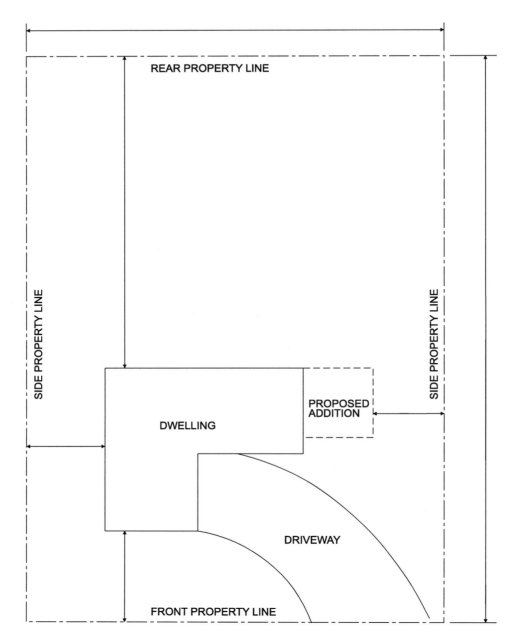

REAR PROPERTY LINE

SIDE PROPERTY LINE

SIDE PROPERTY LINE

DWELLING

PROPOSED
ADDITION

DRIVEWAY

FRONT PROPERTY LINE

STREET

FIGURE 1—TYPICAL SITE PLAN

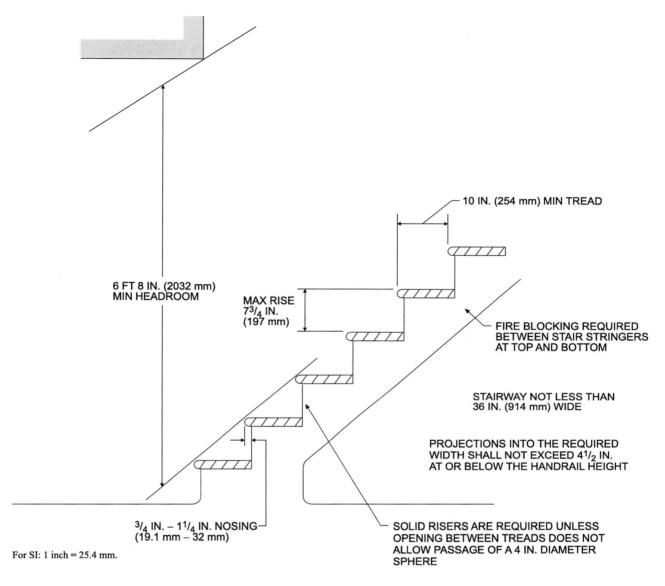

10 IN. (254 mm) MIN TREAD

6 FT 8 IN. (2032 mm) MIN HEADROOM

MAX RISE 7³/₄ IN. (197 mm)

FIRE BLOCKING REQUIRED BETWEEN STAIR STRINGERS AT TOP AND BOTTOM

STAIRWAY NOT LESS THAN 36 IN. (914 mm) WIDE

PROJECTIONS INTO THE REQUIRED WIDTH SHALL NOT EXCEED 4¹/₂ IN. AT OR BELOW THE HANDRAIL HEIGHT

³/₄ IN. – 1¹/₄ IN. NOSING (19.1 mm – 32 mm)

SOLID RISERS ARE REQUIRED UNLESS OPENING BETWEEN TREADS DOES NOT ALLOW PASSAGE OF A 4 IN. DIAMETER SPHERE

For SI: 1 inch = 25.4 mm.

FIGURE 2—STAIR DETAIL

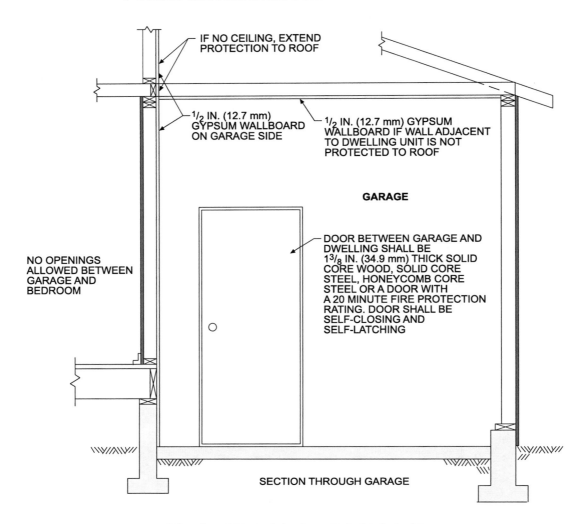

$^5/_8$ IN. (15.9 mm) TYPE X GYPSUM BOARD IS REQUIRED ON CEILING IF THERE IS HABITABLE SPACE ABOVE

IF NO CEILING, EXTEND PROTECTION TO ROOF

$^1/_2$ IN. (12.7 mm) GYPSUM WALLBOARD ON GARAGE SIDE

$^1/_2$ IN. (12.7 mm) GYPSUM WALLBOARD IF WALL ADJACENT TO DWELLING UNIT IS NOT PROTECTED TO ROOF

GARAGE

DOOR BETWEEN GARAGE AND DWELLING SHALL BE $1^3/_8$ IN. (34.9 mm) THICK SOLID CORE WOOD, SOLID CORE STEEL, HONEYCOMB CORE STEEL OR A DOOR WITH A 20 MINUTE FIRE PROTECTION RATING. DOOR SHALL BE SELF-CLOSING AND SELF-LATCHING

NO OPENINGS ALLOWED BETWEEN GARAGE AND BEDROOM

SECTION THROUGH GARAGE

FIGURE 3—SEPARATION OF ATTACHED GARAGE

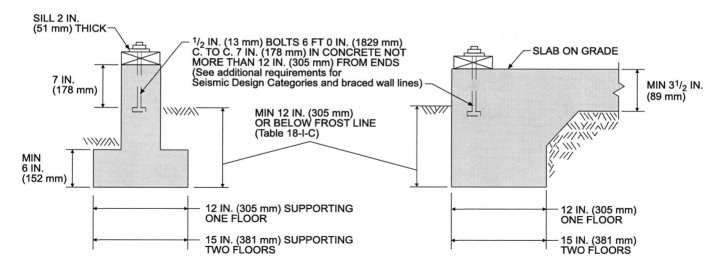

SILL 2 IN. (51 mm) THICK

7 IN. (178 mm)

1/2 IN. (13 mm) BOLTS 6 FT 0 IN. (1829 mm) C. TO C. 7 IN. (178 mm) IN CONCRETE NOT MORE THAN 12 IN. (305 mm) FROM ENDS (See additional requirements for Seismic Design Categories and braced wall lines)

SLAB ON GRADE

MIN 3 1/2 IN. (89 mm)

MIN 12 IN. (305 mm) OR BELOW FROST LINE (Table 18-I-C)

MIN 6 IN. (152 mm)

12 IN. (305 mm) SUPPORTING ONE FLOOR

15 IN. (381 mm) SUPPORTING TWO FLOORS

12 IN. (305 mm) SUPPORTING ONE FLOOR

15 IN. (381 mm) SUPPORTING TWO FLOORS

ALL FOUNDATIONS TO EXTEND INTO NATURAL UNDISTURBED GROUND BELOW FROST LINE

For SI: 1 inch = 25.4 mm.

FIGURE 4—CONCRETE FOUNDATION DETAILS

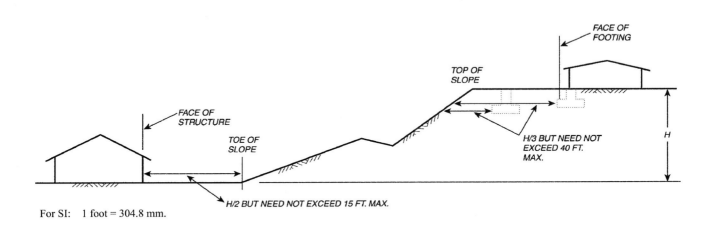

FACE OF FOOTING

TOP OF SLOPE

FACE OF STRUCTURE

TOE OF SLOPE

H/3 BUT NEED NOT EXCEED 40 FT. MAX.

H

H/2 BUT NEED NOT EXCEED 15 FT. MAX.

For SI: 1 foot = 304.8 mm.

FIGURE 5—FOUNDATION CLEARANCES FROM SLOPE

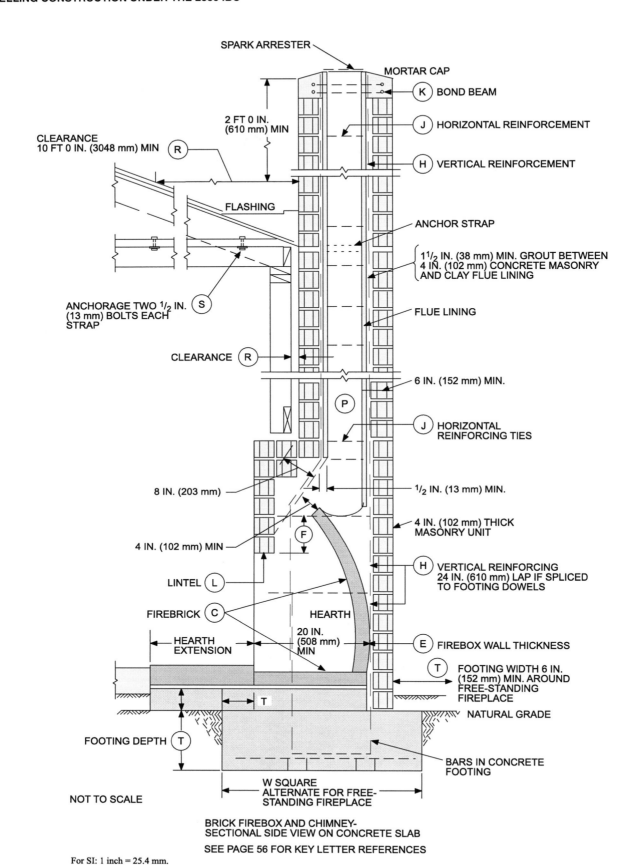

SPARK ARRESTER

MORTAR CAP

(K) BOND BEAM

(J) HORIZONTAL REINFORCEMENT

2 FT 0 IN.
(610 mm) MIN

(H) VERTICAL REINFORCEMENT

CLEARANCE
10 FT 0 IN. (3048 mm) MIN   (R)

FLASHING

ANCHOR STRAP

{ 1¹/₂ IN. (38 mm) MIN. GROUT BETWEEN
4 IN. (102 mm) CONCRETE MASONRY
AND CLAY FLUE LINING

ANCHORAGE TWO ¹/₂ IN.
(13 mm) BOLTS EACH
STRAP   (S)

FLUE LINING

CLEARANCE   (R)

6 IN. (152 mm) MIN.

(P)

(J) HORIZONTAL
REINFORCING TIES

8 IN. (203 mm)

¹/₂ IN. (13 mm) MIN.

4 IN. (102 mm) THICK
MASONRY UNIT

4 IN. (102 mm) MIN.

(F)

LINTEL   (L)

(H) VERTICAL REINFORCING
24 IN. (610 mm) LAP IF SPLICED
TO FOOTING DOWELS

FIREBRICK   (C)

HEARTH

HEARTH
EXTENSION

20 IN.
(508 mm)
MIN

(E) FIREBOX WALL THICKNESS

(T) FOOTING WIDTH 6 IN.
(152 mm) MIN. AROUND
FREE-STANDING
FIREPLACE

NATURAL GRADE

FOOTING DEPTH   (T)

T

BARS IN CONCRETE
FOOTING

NOT TO SCALE

W SQUARE
ALTERNATE FOR FREE-
STANDING FIREPLACE

BRICK FIREBOX AND CHIMNEY-
SECTIONAL SIDE VIEW ON CONCRETE SLAB

SEE PAGE 56 FOR KEY LETTER REFERENCES

For SI: 1 inch = 25.4 mm.

FIGURE 6—TYPICAL MASONRY FIREPLACE AND CHIMNEY

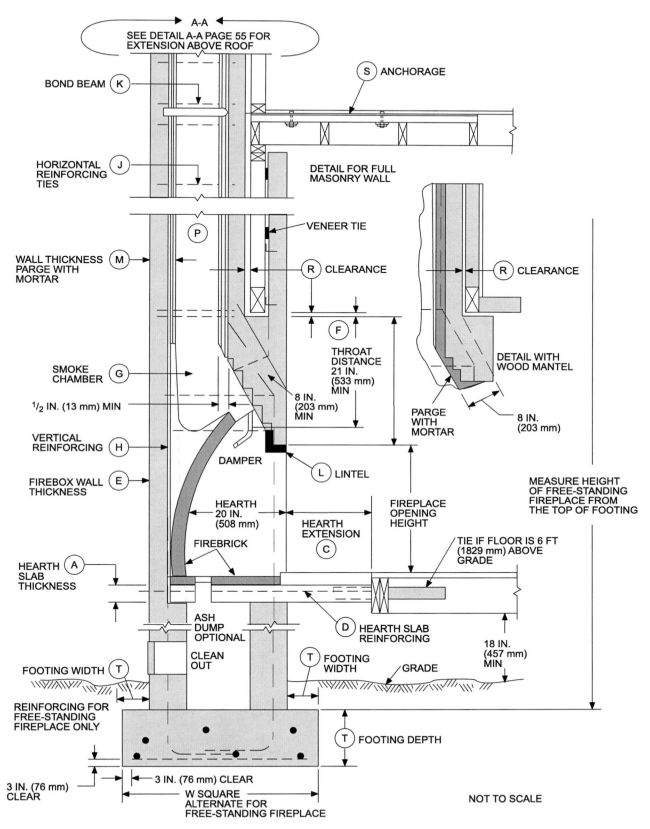

SEE DETAIL A-A PAGE 55 FOR EXTENSION ABOVE ROOF

A-A

BOND BEAM (K)

(S) ANCHORAGE

DETAIL FOR FULL MASONRY WALL

HORIZONTAL REINFORCING TIES (J)

(P)

VENEER TIE

WALL THICKNESS PARGE WITH MORTAR (M)

(R) CLEARANCE

(R) CLEARANCE

(F)

THROAT DISTANCE 21 IN. (533 mm) MIN

DETAIL WITH WOOD MANTEL

SMOKE CHAMBER (G)

8 IN. (203 mm) MIN

1/2 IN. (13 mm) MIN

PARGE WITH MORTAR

8 IN. (203 mm)

VERTICAL REINFORCING (H)

DAMPER

(L) LINTEL

FIREBOX WALL THICKNESS (E)

MEASURE HEIGHT OF FREE-STANDING FIREPLACE FROM THE TOP OF FOOTING

HEARTH 20 IN. (508 mm)

HEARTH EXTENSION (C)

FIREPLACE OPENING HEIGHT

FIREBRICK

TIE IF FLOOR IS 6 FT (1829 mm) ABOVE GRADE

HEARTH SLAB THICKNESS (A)

ASH DUMP OPTIONAL

(D) HEARTH SLAB REINFORCING

CLEAN OUT

FOOTING WIDTH (T)

(T) FOOTING WIDTH

GRADE

18 IN. (457 mm) MIN

REINFORCING FOR FREE-STANDING FIREPLACE ONLY

(T) FOOTING DEPTH

3 IN. (76 mm) CLEAR

3 IN. (76 mm) CLEAR

W SQUARE ALTERNATE FOR FREE-STANDING FIREPLACE

NOT TO SCALE

For SI: 1 inch = 25.4 mm.

BRICK FIREBOX AND CHIMNEY SECTIONAL SIDE VIEW ON WOOD FLOOR

FIGURE 6—TYPICAL MASONRY FIREPLACE AND CHIMNEY—(continued)

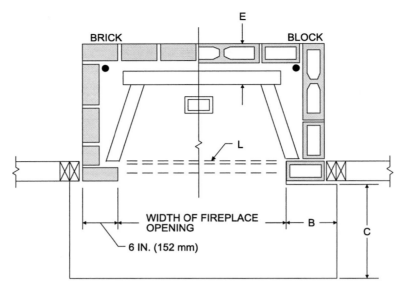

PLAN VIEW

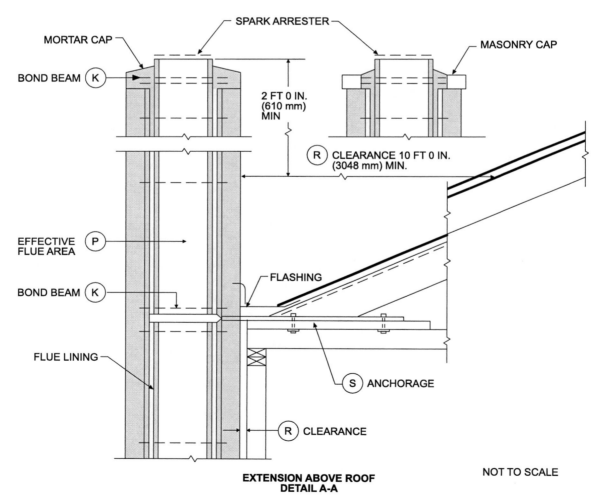

**EXTENSION ABOVE ROOF
DETAIL A-A**

NOT TO SCALE

For SI: 1 inch = 25.4 mm, 1 foot = 304.8 mm.

FIGURE 6—TYPICAL MASONRY FIREPLACE AND CHIMNEY—(continued)

SUMMARY OF REQUIREMENTS FOR MASONRY FIREPLACES AND CHIMNEYS

ITEM	LETTER[a]	REQUIREMENTS
Hearth slab thickness	A	4″
Hearth extension (each side of opening)	B	8″ fireplace opening < 6 square foot. 12″ fireplace opening ≥ 6 square foot.
Hearth extension (front of opening)	C	16″ fireplace opening < 6 square foot. 20″ fireplace opening ≥ 6 square foot.
Hearth slab reinforcing	D	Reinforced to carry its own weight and all imposed loads.
Thickness of wall of firebox	E	10″ solid brick or 8″ where a firebrick lining is used. Joints in firebrick $^1/_4$″ maximum.
Distance from top of opening to throat	F	8″
Smoke chamber wall thickness Unlined walls	G	6″ 8″
Chimney Vertical reinforcing[a]	H	Four No. 4 full-length bars for chimney up to 40″ wide. Add two No. 4 bars for each additional 40″ or fraction of width or each additional flue.
Horizontal reinforcing	J	$^1/_4$″ ties at 18″ and two ties at each bend in vertical steel.
Bond beams	K	No specified requirements.
Fireplace lintel	L	Noncombustible material.
Chimney walls with flue lining	M	Solid masonry units or hollow masonry units grouted solid with at least 4 inch nominal thickness.
Distances between adjacent flues	—	See Section 2113.14
Effective flue area (based on area of fireplace opening)	P	See Section 2113.16
Clearances: Combustible material Mantel and trim Above roof	R	See Sections 2113.19, 2111.11 See Section 2111.11 3′ at roofline and 2′ at 10′.
Anchorage[a] Strap Number Embedment into chimney Fasten to Bolts	S	$^3/_{16}$″ × 1″ Two 12″ hooked around outer bar with 6″ extension. 4 joists Two $^1/_2$″ diameter.
Footing Thickness Width	T	12″ min. 6″ each side of fireplace wall.

For SI: 1 inch = 25.4 mm, 1 foot = 304.8 mm, 1 square foot = 0.0929 m².
a. Not required in Seismic Design Category A, B or C.

FIGURE 6—TYPICAL MASONRY FIREPLACE AND CHIMNEY—(continued)

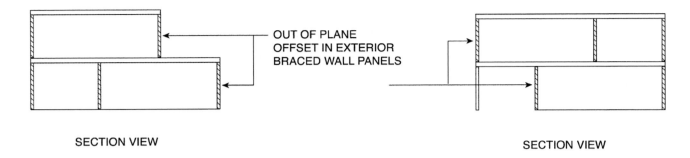

OUT OF PLANE
OFFSET IN EXTERIOR
BRACED WALL PANELS

SECTION VIEW   SECTION VIEW

BRACED WALL PANELS OUT OF PLANE

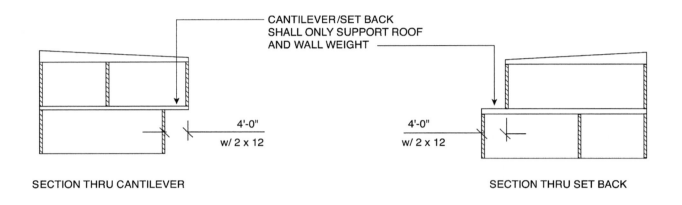

CANTILEVER/SET BACK
SHALL ONLY SUPPORT ROOF
AND WALL WEIGHT

4'-0"
w/ 2 x 12

4'-0"
w/ 2 x 12

SECTION THRU CANTILEVER   SECTION THRU SET BACK

BRACED WALL PANELS SUPPORTED BY CANTILEVER OR SET BACK

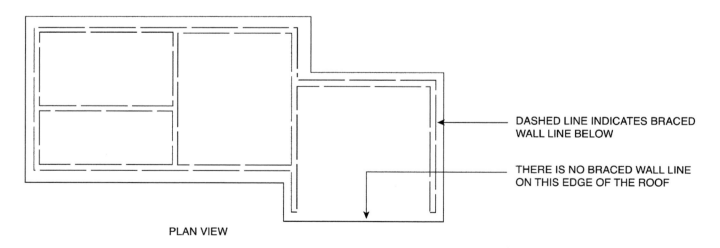

DASHED LINE INDICATES BRACED
WALL LINE BELOW

THERE IS NO BRACED WALL LINE
ON THIS EDGE OF THE ROOF

PLAN VIEW

FLOOR OR ROOF NOT SUPPORTED ON ALL EDGES

FIGURE 7—IRREGULAR STRUCTURES IN SEISMIC DESIGN CATEGORY D OR E (Section 2308.12.6)

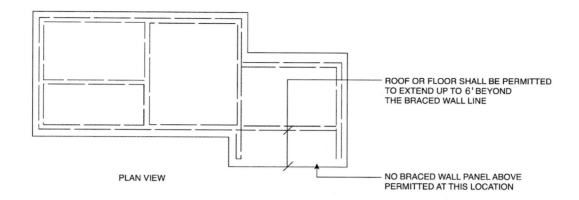

PLAN VIEW

ROOF OR FLOOR SHALL BE PERMITTED TO EXTEND UP TO 6' BEYOND THE BRACED WALL LINE

NO BRACED WALL PANEL ABOVE PERMITTED AT THIS LOCATION

ROOF OR FLOOR EXTENSION BEYOND BRACED WALL LINE

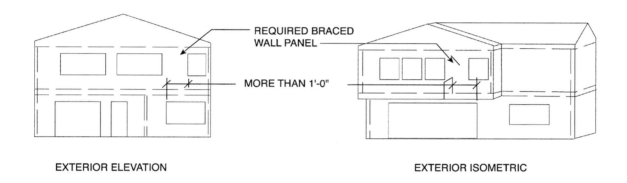

REQUIRED BRACED WALL PANEL

MORE THAN 1'-0"

EXTERIOR ELEVATION

EXTERIOR ISOMETRIC

BRACED WALL PANEL EXTENSION OVER OPENING

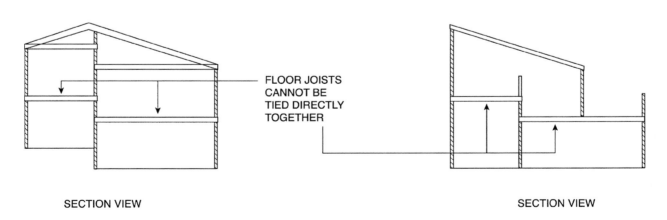

FLOOR JOISTS CANNOT BE TIED DIRECTLY TOGETHER

SECTION VIEW

SECTION VIEW

PORTIONS OF FLOOR LEVEL OFFSET VERTICALLY

FIGURE 7—IRREGULAR STRUCTURES IN SEISMIC DESIGN CATEGORY D OR E—(continued)

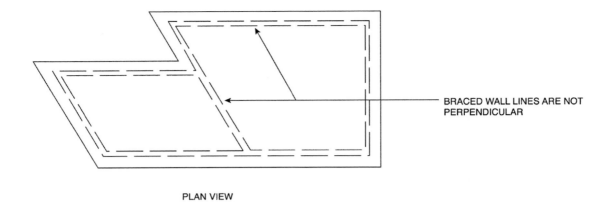

PLAN VIEW

BRACED WALL LINES NOT PERPENDICULAR

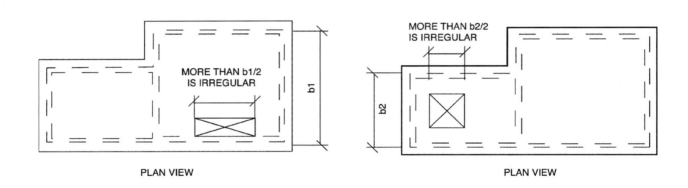

OPENING LIMITATIONS FOR FLOOR AND ROOF DIAPHRAGMS

FIGURE 7—IRREGULAR STRUCTURES IN SEISMIC DESIGN CATEGORY D OR E—(continued)

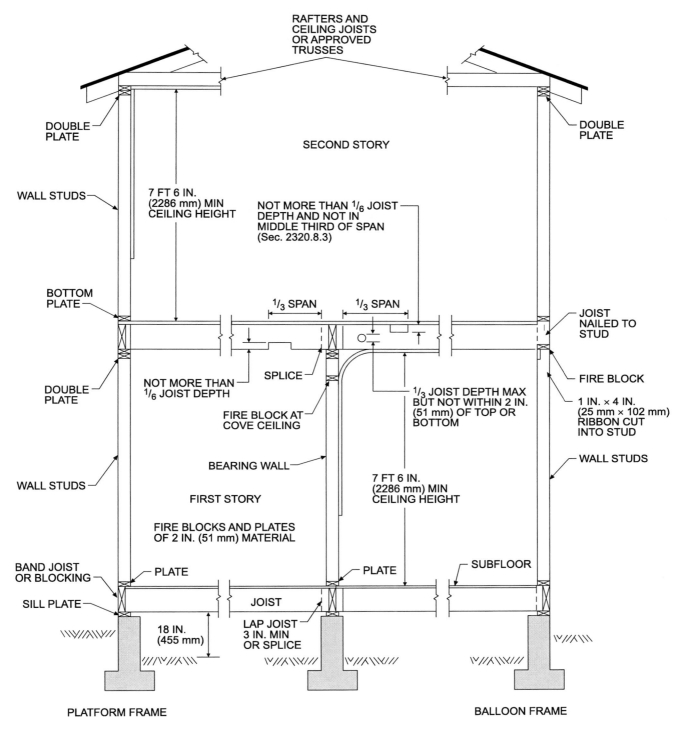

RAFTERS AND
CEILING JOISTS
OR APPROVED
TRUSSES

DOUBLE
PLATE

DOUBLE
PLATE

SECOND STORY

WALL STUDS

7 FT 6 IN.
(2286 mm) MIN
CEILING HEIGHT

NOT MORE THAN 1/6 JOIST
DEPTH AND NOT IN
MIDDLE THIRD OF SPAN
(Sec. 2320.8.3)

BOTTOM
PLATE

1/3 SPAN   1/3 SPAN

JOIST
NAILED TO
STUD

DOUBLE
PLATE

SPLICE

1/3 JOIST DEPTH MAX
BUT NOT WITHIN 2 IN.
(51 mm) OF TOP OR
BOTTOM

FIRE BLOCK

NOT MORE THAN
1/6 JOIST DEPTH

1 IN. × 4 IN.
(25 mm × 102 mm)
RIBBON CUT
INTO STUD

FIRE BLOCK AT
COVE CEILING

BEARING WALL

WALL STUDS

WALL STUDS

FIRST STORY

7 FT 6 IN.
(2286 mm) MIN
CEILING HEIGHT

FIRE BLOCKS AND PLATES
OF 2 IN. (51 mm) MATERIAL

BAND JOIST
OR BLOCKING

PLATE

PLATE

SUBFLOOR

SILL PLATE

JOIST

18 IN.
(455 mm)

LAP JOIST
3 IN. MIN
OR SPLICE

PLATFORM FRAME

BALLOON FRAME

For SI: 1 inch = 25.4 mm.

FIGURE 8—FRAMING SECTION

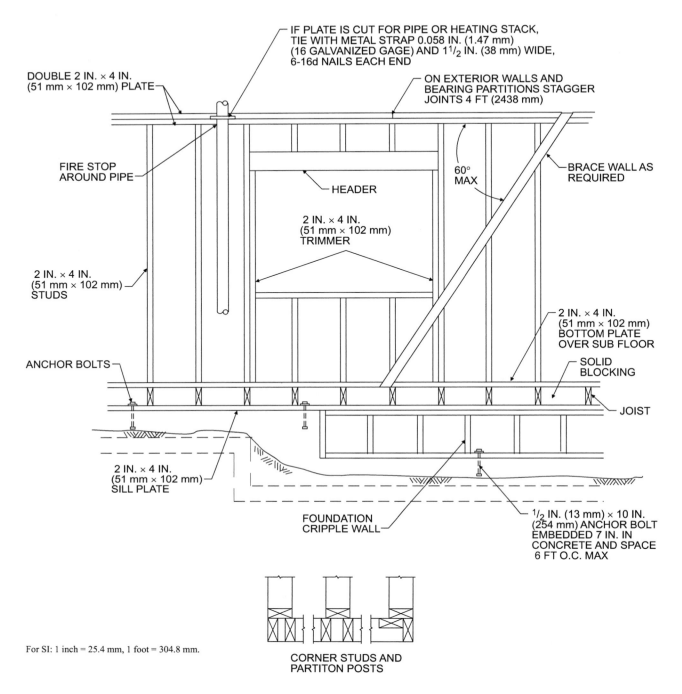

IF PLATE IS CUT FOR PIPE OR HEATING STACK, TIE WITH METAL STRAP 0.058 IN. (1.47 mm) (16 GALVANIZED GAGE) AND 1 1/2 IN. (38 mm) WIDE, 6-16d NAILS EACH END

DOUBLE 2 IN. × 4 IN. (51 mm × 102 mm) PLATE

ON EXTERIOR WALLS AND BEARING PARTITIONS STAGGER JOINTS 4 FT (2438 mm)

FIRE STOP AROUND PIPE

HEADER

60° MAX

BRACE WALL AS REQUIRED

2 IN. × 4 IN. (51 mm × 102 mm) TRIMMER

2 IN. × 4 IN. (51 mm × 102 mm) STUDS

2 IN. × 4 IN. (51 mm × 102 mm) BOTTOM PLATE OVER SUB FLOOR

ANCHOR BOLTS

SOLID BLOCKING

JOIST

2 IN. × 4 IN. (51 mm × 102 mm) SILL PLATE

FOUNDATION CRIPPLE WALL

1/2 IN. (13 mm) × 10 IN. (254 mm) ANCHOR BOLT EMBEDDED 7 IN. IN CONCRETE AND SPACE 6 FT O.C. MAX

For SI: 1 inch = 25.4 mm, 1 foot = 304.8 mm.

CORNER STUDS AND PARTITON POSTS

FIGURE 9—WALL FRAMING DETAILS

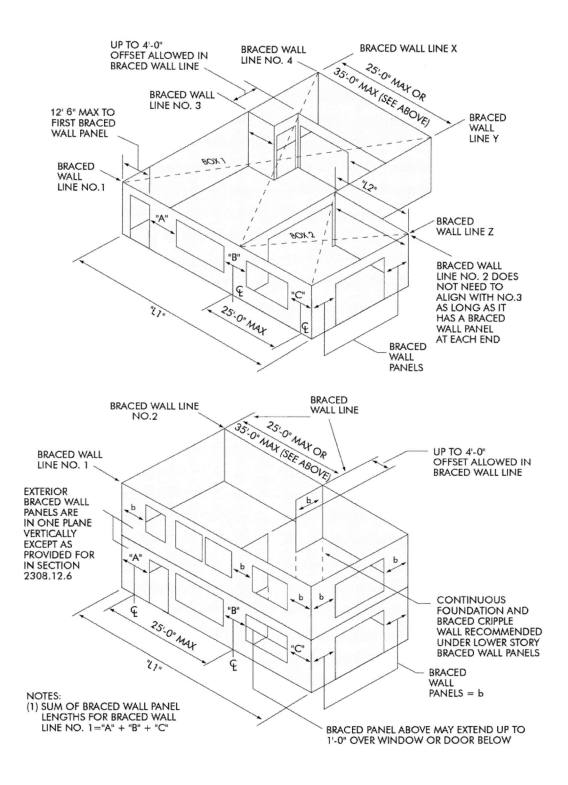

FIGURE 10—BASIC COMPONENTS OF THE LATERAL BRACING SYSTEM

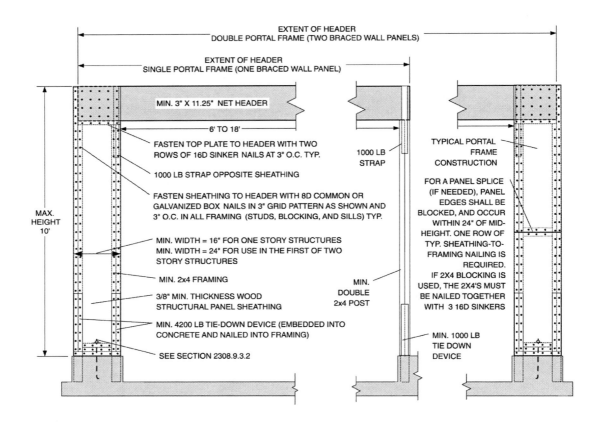

EXTENT OF HEADER
DOUBLE PORTAL FRAME (TWO BRACED WALL PANELS)

EXTENT OF HEADER
SINGLE PORTAL FRAME (ONE BRACED WALL PANEL)

MIN. 3" X 11.25" NET HEADER

6' TO 18'

FASTEN TOP PLATE TO HEADER WITH TWO
ROWS OF 16D SINKER NAILS AT 3" O.C. TYP.

1000 LB STRAP OPPOSITE SHEATHING

FASTEN SHEATHING TO HEADER WITH 8D COMMON OR
GALVANIZED BOX NAILS IN 3" GRID PATTERN AS SHOWN AND
3" O.C. IN ALL FRAMING (STUDS, BLOCKING, AND SILLS) TYP.

MIN. WIDTH = 16" FOR ONE STORY STRUCTURES
MIN. WIDTH = 24" FOR USE IN THE FIRST OF TWO
STORY STRUCTURES

MIN. 2x4 FRAMING

3/8" MIN. THICKNESS WOOD
STRUCTURAL PANEL SHEATHING

MIN. 4200 LB TIE-DOWN DEVICE (EMBEDDED INTO
CONCRETE AND NAILED INTO FRAMING)

SEE SECTION 2308.9.3.2

MAX.
HEIGHT
10'

1000 LB
STRAP

MIN.
DOUBLE
2x4 POST

MIN. 1000 LB
TIE DOWN
DEVICE

TYPICAL PORTAL
FRAME
CONSTRUCTION

FOR A PANEL SPLICE
(IF NEEDED), PANEL
EDGES SHALL BE
BLOCKED, AND OCCUR
WITHIN 24" OF MID-
HEIGHT. ONE ROW OF
TYP. SHEATHING-TO-
FRAMING NAILING IS
REQUIRED.
IF 2X4 BLOCKING IS
USED, THE 2X4'S MUST
BE NAILED TOGETHER
WITH 3 16D SINKERS

FIGURE 11—ALTERNATE BRACED WALL PANEL ADJACENT TO A DOOR OR WINDOW OPENING

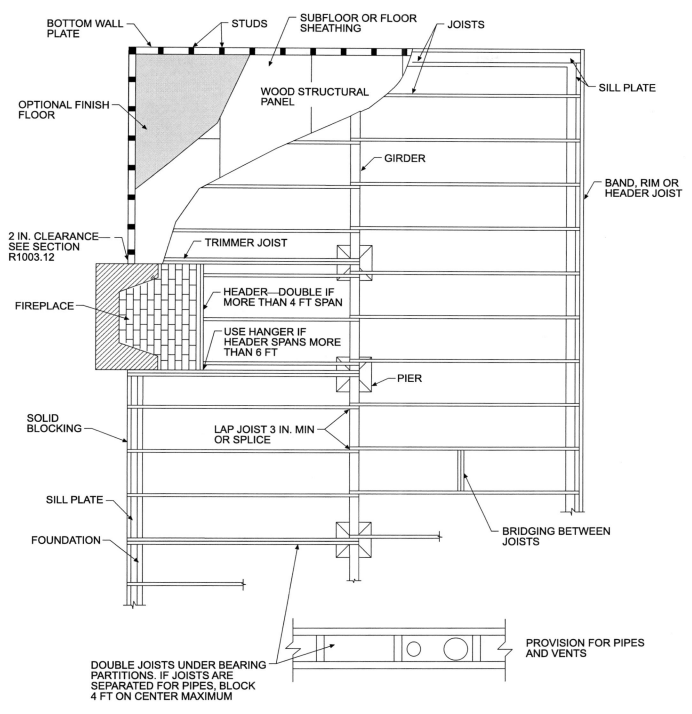

BOTTOM WALL PLATE

STUDS

SUBFLOOR OR FLOOR SHEATHING

JOISTS

OPTIONAL FINISH FLOOR

WOOD STRUCTURAL PANEL

SILL PLATE

GIRDER

BAND, RIM OR HEADER JOIST

2 IN. CLEARANCE SEE SECTION R1003.12

TRIMMER JOIST

FIREPLACE

HEADER—DOUBLE IF MORE THAN 4 FT SPAN

USE HANGER IF HEADER SPANS MORE THAN 6 FT

PIER

SOLID BLOCKING

LAP JOIST 3 IN. MIN OR SPLICE

SILL PLATE

FOUNDATION

BRIDGING BETWEEN JOISTS

DOUBLE JOISTS UNDER BEARING PARTITIONS. IF JOISTS ARE SEPARATED FOR PIPES, BLOCK 4 FT ON CENTER MAXIMUM

PROVISION FOR PIPES AND VENTS

For SI: 1 inch = 25.4 mm, 1 foot = 304.8 mm.

FIGURE 12—FLOOR CONSTRUCTION DETAIL

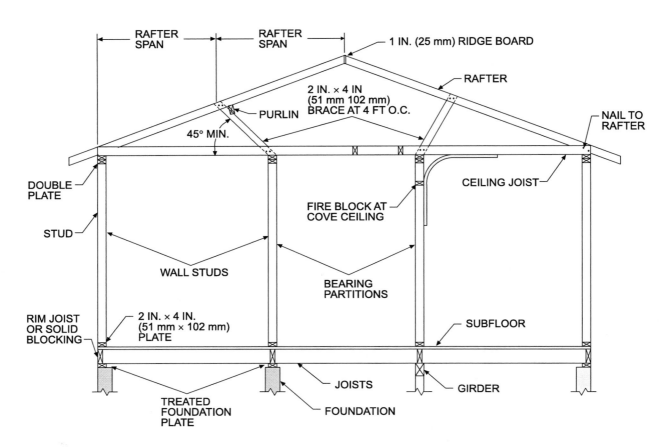

FIGURE 13—BRACED RAFTER CONSTRUCTION

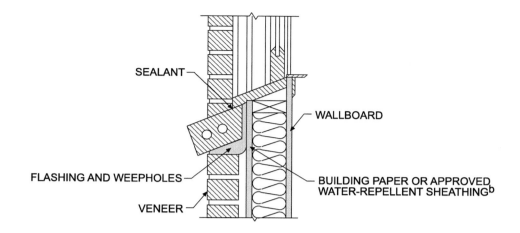

SEALANT

WALLBOARD

FLASHING AND WEEPHOLES

BUILDING PAPER OR APPROVED WATER-REPELLENT SHEATHING[b]

VENEER

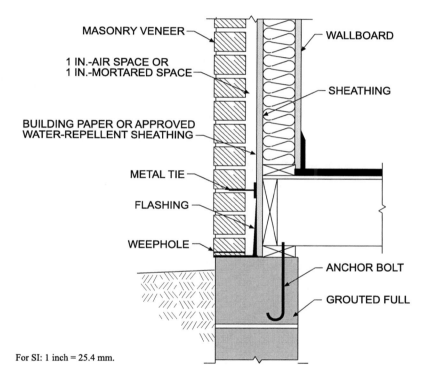

MASONRY VENEER

WALLBOARD

1 IN.-AIR SPACE OR
1 IN.-MORTARED SPACE

BUILDING PAPER OR APPROVED
WATER-REPELLENT SHEATHING

SHEATHING

METAL TIE

FLASHING

WEEPHOLE

ANCHOR BOLT

GROUTED FULL

For SI: 1 inch = 25.4 mm.

FIGURE 14—MASONRY VENEER WALL DETAILS

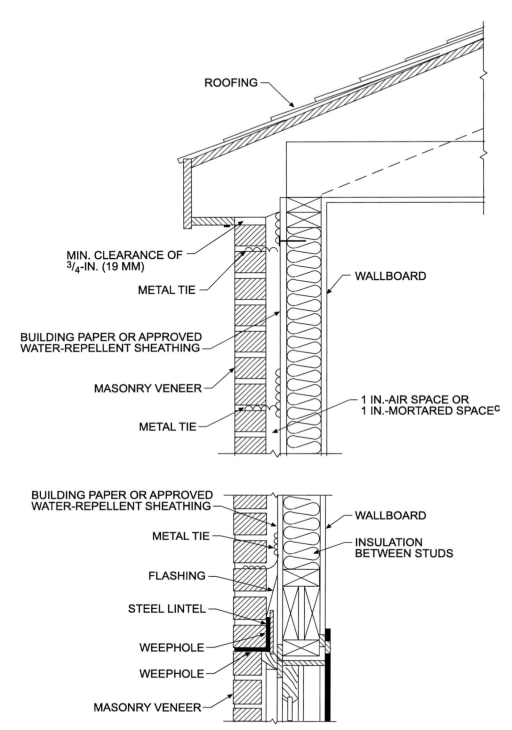

ROOFING

MIN. CLEARANCE OF ³/₄-IN. (19 MM)

METAL TIE

WALLBOARD

BUILDING PAPER OR APPROVED WATER-REPELLENT SHEATHING

MASONRY VENEER

METAL TIE

1 IN.-AIR SPACE OR 1 IN.-MORTARED SPACE[c]

BUILDING PAPER OR APPROVED WATER-REPELLENT SHEATHING

METAL TIE

WALLBOARD

INSULATION BETWEEN STUDS

FLASHING

STEEL LINTEL

WEEPHOLE

WEEPHOLE

MASONRY VENEER

For SI: 1 inch = 25.4 mm.

FIGURE 14—MASONRY VENEER WALL DETAILS—(continued)

Membership Application

This form may be photocopied

MEMBERSHIP CATEGORIES AND DUES*— ANNUAL MEMBERSHIP

Special membership structures are also available for Educational Institutions and Federal Agencies.
For more information, please visit www.iccsafe.org/membership or call 1-888-ICC-SAFE (422-7233), x33804.

GOVERNMENTAL MEMBER**

Government/Municipality (including agencies, departments or units) engaged in administration, formulation or enforcement of laws, regulations or ordinances relating to public health, safety and welfare. Annual member dues (by population) are shown below. Please verify the current ICC membership status of your employer prior to applying.

☐ Up to 50,000..........$100   ☐ 50,001–150,000........ $180   ☐ 150,001+.......... $280

**A Governmental Member may designate 4 to 12 voting representatives (based on population) who are employees or officials of that governmental member and are actively engaged on a full- or part-time basis in the administration, formulation or enforcement of laws, regulations or ordinances relating to public health, safety and welfare. Number of representatives is based on population. All dues for representatives have been included in the annual member dues payment. Please call 1-888-ICC-SAFE (422-7233), x33804 for information about how to designate your voting representatives.

☐ **CORPORATE MEMBERS ($300)**   An association, society, testing laboratory, manufacturer, company or corporation

INDIVIDUAL MEMBERS

☐ **PROFESSIONAL ($150)**   A design professional duly licensed or registered by any state or other recognized governmental agency

☐ **COOPERATING ($150)**   An individual who is interested in International Code Council purposes and objectives and would like to take advantage of membership benefits

☐ **CERTIFIED ($75)**   An individual who holds a current Legacy or International Code Council certification

☐ **ASSOCIATE ($35)**   An employee of a current ICC Governmental Member

☐ **STUDENT ($25)**   An individual who is enrolled in classes or a course of study including at least 12 hours of classroom instruction per week

☐ **RETIRED ($20)**   A former governmental representative, corporate or individual member who has retired

☐ **TEMPORARY ($50)**   A trial 4-month non-renewing membership with all of the benefits provided to our individuals members, except a complimentary code book.

New Governmental and Corporate Members will receive a free package of 7 code books. New Individual Members will receive one free code book. Upon receipt of your completed application and payment, you will be contacted by an ICC Member Services Representative regarding your free code package or code book. For more information, please visit www.iccsafe.org/membership or call 1-888-ICC-SAFE (422-7233), x33804.

Please print clearly or type information below:

Name
_____

Name of Jurisdiction, Association, Institute or Company, etc.
_____

Title
_____

Billing Address
_____

City _____ State _____ Zip+4 _____

Street Address for Shipping
_____

City _____ State _____ Zip+4 _____

E-mail
_____

Telephone
_____

Tax Exempt Number (If applicable, must attach copy of tax exempt license if claiming an exemption)
_____

Payment Information:

VISA, MC, AMEX or DISCOVER Account Number   Exp. Date

Return this application to:
International Code Council
Attn: Membership
5360 Workman Mill Road
Whittier, CA 90601-2298

Toll Free: 1-888-ICC-SAFE (1-888-422-7233), x33804
FAX: (562) 692-6031 (Los Angeles District Office)
Or, apply online at **www.iccsafe.org/membership**.
Please refer to Tracking Number 66-07-027 when applying.

If you have any questions about membership in the International Code Council, call 1-888-ICC-SAFE (1-888-422-7233), x33804 and request a Member Services Representative.

REF 66-07-027

*Membership categories and dues subject to change.
Please visit www.iccsafe.org/membership for the most current information.